Marnie Sinclair

Balance & Imbalance

A Celebration of Nature and a Call to Action

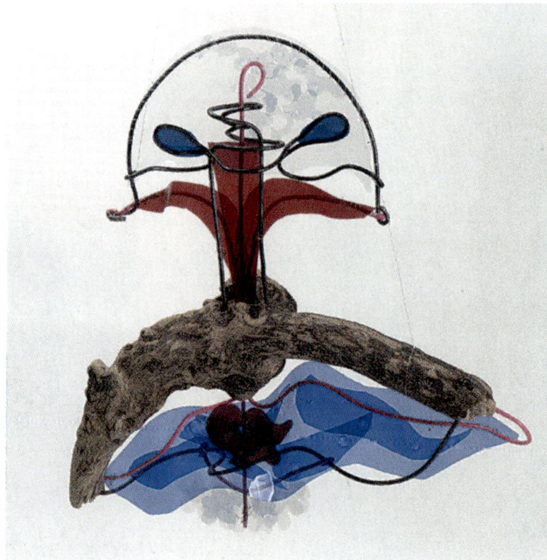

Dedicated to my grandchildren, August, Winnie, Lucy, Abe, and Lila, who drive my passion and make my heart sing.

ISBN: 978-0-692-86395-4
Printed in the United States of America.

WHY

Last year *I was shocked to see my first and only monarch butterfly standing motionless with torn wings on the hot pavement of a store parking lot. I stopped to study it momentarily and then went on my way, thinking, "Too bad, but so it goes, survival of the fittest." But by the time I got to the door I was feeling really badly, and I turned around to see if it was still there. It was, a tiny creature surrounded by huge cars and human activity. I walked back over to it, knelt down and extended my finger. To my surprise, without hesitation it climbed on. I carried it over to the edge of the pavement and set it down in the cool moist shade of the grass. Sadly, as I left, I wished it well, with the hope that it could die with some comfort and peace.*

The story of my encounter with the dying monarch explains why I have written this book: Every form of life is interdependent on all other forms of life that help to sustain Nature's vast diversity. A state of balance is the criterion I use to understand the complex web of life that exists all around us on this beautiful planet. We all fit together with a precision and balance that is second to none. There is no excess or waste. Most creatures take only what they need to survive. (When was the last time you saw a fat deer?) In spite of how sophisticated we think we are, and even though we are 7+ billion strong, we are still only one piece of a complicated interdependent puzzle that makes up our very existence on Earth.

Since I was a small child I have been fascinated and inspired by how inventive and magical Nature is in assuring the survival of each species, whether through patterns, shapes, camouflage, or cleverness. The iconic honeybee transfers pollen from plant to plant, assuring the continuation of a huge number of plants and edible crops that sustain us. The magical metamorphosis of a chubby, gushy caterpillar, chomping its way across a leaf, into a magnificent, beautifully colored butterfly that can spread its wings and fly, still seems like a miracle to me! Everywhere in Nature you see this amazing interconnection that is full of harmony and balance. Witness the huge moose that tramples and beds down on shrubs and grasses, and in the process, creates layers of organic material that become prime habitats and sources of food for birds and insects. Our dependence on each other is never-ending and life-giving.

I'm an artist and a grandmother, and my views and passions come from that place. And so it was that in 2010 I set out to try and make sense of climate change. I decided to make one piece of abstract art a week for 52 weeks that related directly to environmental degradation issues and more specifically, to climate change. I researched the science behind each story and presented the art and the story together.

Perhaps my favorite piece of abstract art is "Earth Nest." *(See page 4.)* During a visit with my infant granddaughter Winnie I was struck by the extraordinary innocence, vulnerability, and wonderment that is ever present in the faces of babies of all species. When they are threatened, there is nothing that will get you out of a left-brained mode faster than their safety. We are instantly removed from our intellectual perch and thrown into a fierce protective place that reminds us of our animal roots. It occurred to me that as the fossil fuel industry continues to dominate the energy needs of most of the industrial nations, with the non-industrial countries in fast pursuit, we are soiling our only nest. The children of these many nations will suffer the consequences of the excess carbon dioxide (CO2) in the atmosphere, which is rapidly heating up the planet. My recognition of this threat, and of the strength of the maternal instinct inspired me to create "Earth Nest."

Because the subject of climate change is so sweeping and so complicated, it was important for my sense of balance and well-being to make the art as aesthetic and pleasing as possible. Making so many pieces in such a short time was intense and amazing. I was in the "flow," which meant I never stopped thinking about the creative process and the next piece of art. What really surprised me was how the muse was ever-present and calling the shots. I had always thought it only happened when you were physically making the work, but now I know that the inner voice is strong even as the creative ideas are being formulated in your mind. I made it through the first year using the playfulness of art as a counterbalance to the scary news, and as a cushion for the viewer from the true horrors we are inflicting on our only home. At the end of that year I had 40 pieces of art that were specifically about climate change, so I decided to make a movie that used art to illustrate the story of climate change. The movie includes interviews with three experts in the field: Bill McKibben, founder of 350.org, George Woodwell, chief scientist of Woods Hole Oceanographic Institute and the founder of the Woods Hole Research Center, and Jerry Sullivan, a retired NOAA physicist, whose expertise was in the satellite study of temperature in the troposphere. Their stories about the state of our planet are straight forward, scientifically based, and true. My movie, "Nature's Spin Through Art," can be seen on YouTube. *(All sources and links are listed at the back of this book.)*

There are lots more details to the climate change story, but I have to say in doing the research I felt such despair that I found I sometimes couldn't go on. I needed balance, so I decided to focus on the magic in Nature that had moved me so much as a child. I became fascinated with how plants communicate with each other and began to investigate their stories. My visual interpretation of the stories I discovered inspired me to look further. My recent artwork reflects the extraordinary diversity in Nature that ensures each species' survival. I have loved discovering interspecies symbiosis, adaptation, and communication.

This book shows the progression of my understanding of climate change. It begins with the wonder of Nature's balance, it explores some ways we have negatively impacted specific species, it describes how climate change is degrading the planet's environment, and it ends with a celebration of Nature's ability to adapt to the changes we have imposed upon the planet. I want this book to challenge, enlighten, and inspire you to action. We are surrounded by the extraordinary natural world that inspires me every day and gives me confidence in our future. I hope my artwork brings you inspiration and hope as well.

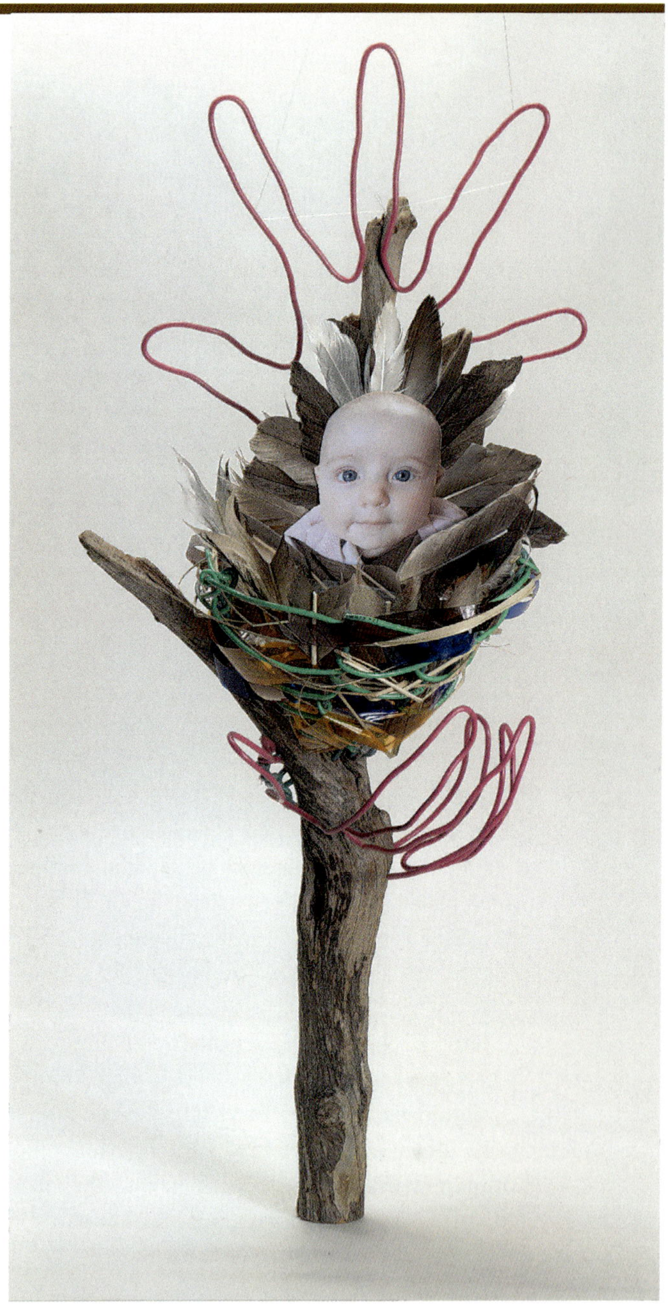

"Earth Nest" shows the author's grandchild, safe in a nest. The wire hand behind the child's head demands a stop to the destruction and the other hand offers a cradle for protection. (wire, wood, photograph, feathers)

BALANCE

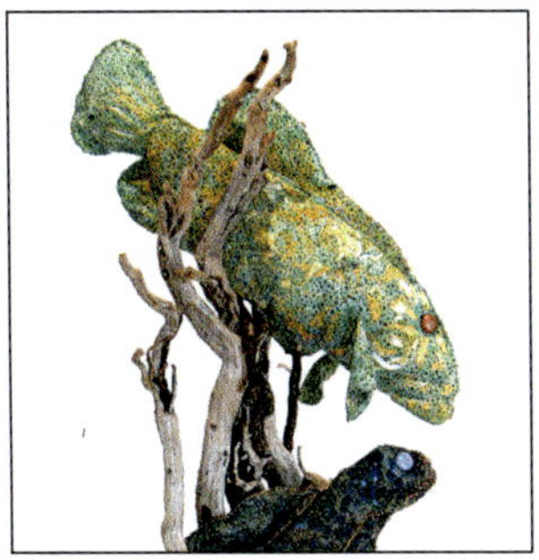

Nature is perfectly balanced. It never fails to leave me in awe when I witness the complexity and intelligence that exists within each species and the ways species interact. Each plant and animal species is part of a large interactive web that maintains a perfect balance of food, water, and shelter, to assure a sustainable existence.

Interspecies symbiotic relationships are exquisitely balanced as plants and animals work to each others' benefit to survive. Recently plants have been found to have amazing abilities to communicate with each other through pheromones, their root systems, and their fluids. These communications protect the plants from predators, allow them to recognize and protect kin, and to produce nutrients that benefit both the host plant and its neighbors. Nature's timing is perfectly balanced, as illustrated by migrations that are triggered by the changing weather, or by baby birds that hatch in time to coordinate with the hatching of the insects that will become their first food. Another example of good timing and perfect balance is the survival of the moose on the Kenai Peninsula in Alaska. The moose all give birth to their calves on the same day in March every year. The calving coincides with the reappearance of the grizzly bears emerging from hibernation. The bears are extremely hungry and the moose calves are easy prey. But because the peninsula is abundant with moose calves, the grizzlies are able to eat and survive without wiping out the new generation of moose. This example and so many others remind us of the balance that is so extraordinary in Nature.

THE PITCHER AND BAT

The pitcher plant from Borneo has developed a unique symbiotic partnership with bats. The carnivorous plant attracts bats, not to eat them but to ingest the nutrients found in their guano. The bat in exchange is provided with a safe cozy place to rest. This give-and-take arrangement is not the only interesting aspect of their relationship: Unlike other plants that send out pheromones to communicate, this pitcher plant is the only plant that uses echo location to guide bats to it, similar to the sonar the bats use to navigate.

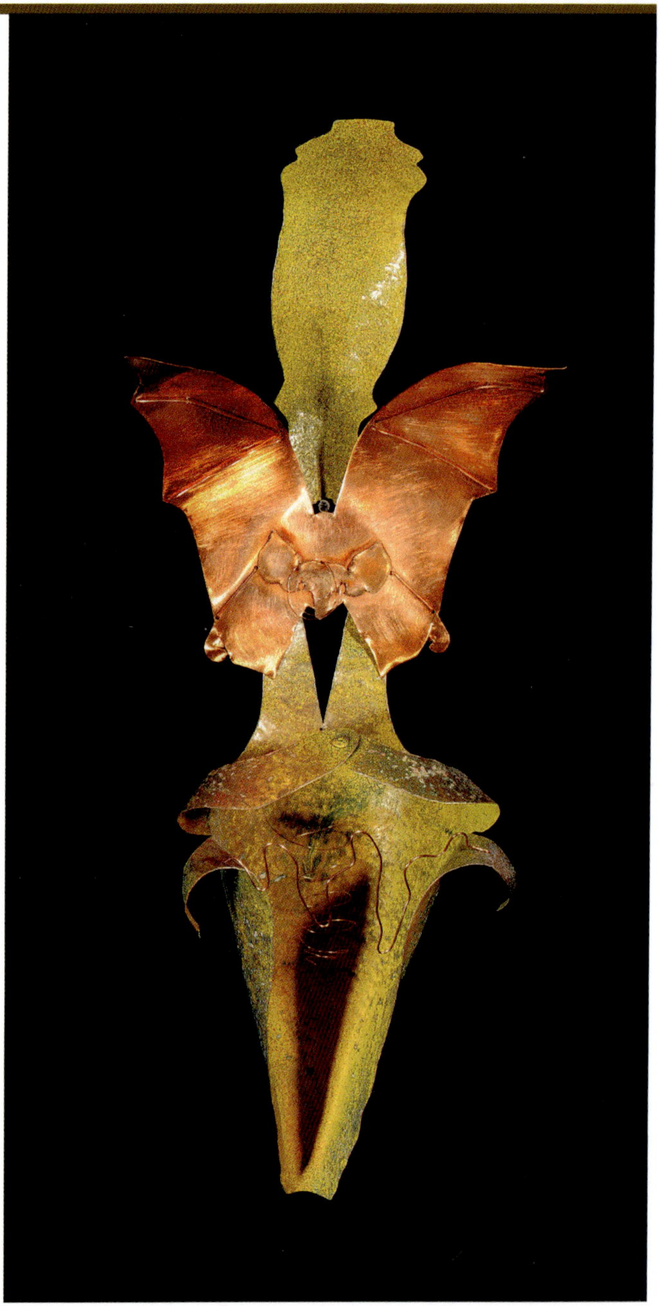

"The Pitcher and Bat" shows the pitcher plant in the background, and the bat, with outspread wings, is poised over the cup of the plant's blossom. The bat's guano is represented by draped wire on the lip of the blossom, and thin wire wound in spirals represents the sonar connection between the plant and the bat. (metal, wire)

Wild tobacco

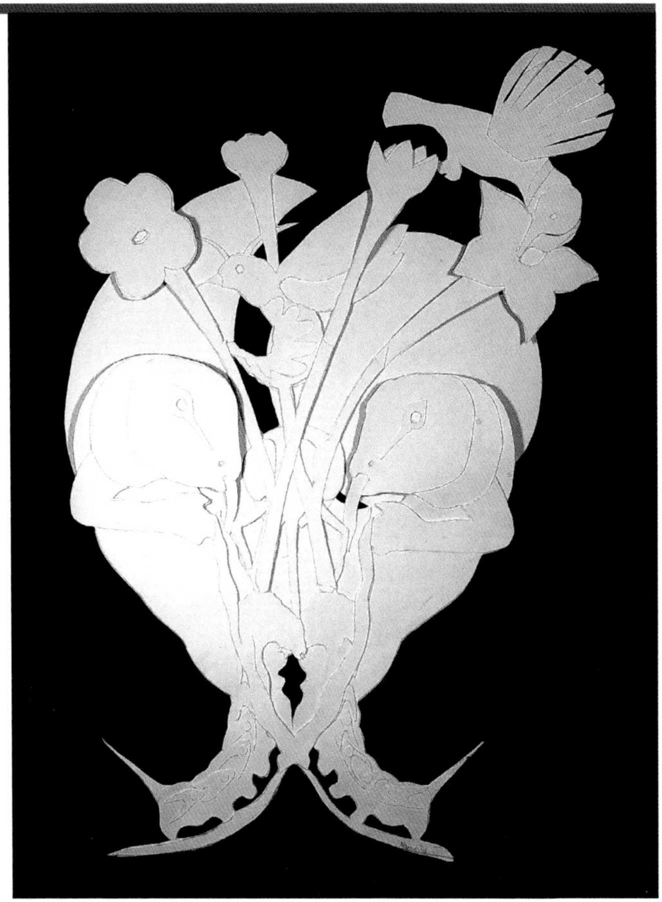

"Wild Tobacco" illustrates the various creatures that interact with the wild tobacco plant. On the top right is a hummingbird feeding from a blossom. In the middle of the collage is a big-eyed bug, flanked by two lizards about to attack the two hornworm caterpillars at the base of the plant. (paper)

The seeds of the Wild Tobacco plant only germinate after a fire, and the seeds can remain viable for hundreds of years, waiting for a fire to occur. After such a long wait, the plant defends itself vigorously against predators and has several techniques to insure it is in balance with its environment. If attacked by insect pests, the plant releases nicotine through its leaves. The nicotine is a neurotoxin which kills all of the insects except for the hornworm caterpillar that is immune to it. These caterpillars become hawk moths, the plant's nocturnal pollinator. If too many hawk moth eggs hatch, threatening the plant's survival, the wild tobacco plant produces a sweet nectar that the baby caterpillars can't resist. This nectar gives the tiny creatures a body odor that is irresistible to big-eyed bugs that flock to the plant to eat them. If the big-eyed bugs don't eat enough of the voracious caterpillars, the plant releases a pheromone that attracts a ground-foraging lizard that feasts on the caterpillars. If all of these steps still don't restore the natural balance to the tobacco plant, it then changes from a nocturnal plant to a diurnal (daytime) plant. The daytime version of the wild tobacco plant has a completely different structure to its flowers as well as a change to the sugar content of its nectar, attracting hummingbirds as daytime pollinators. This amazing transformation from a nocturnal plant to a daytime plant happens in just eight days.

Air and Ant

Epiphytes (air plants) have a unique symbiotic relationship with ants. Their hollow stems, inner chambers, and succulent leaves are used as living quarters for whole colonies of ants. The ants live in the smooth-walled interior chambers, while the outer chambers with rough walls are used to store ant waste and dead ants. The plant is able to access needed nutrients from the ant waste through the rough walls, and in exchange, the ants have a safe place to live. Some species of air plants also provide nectar specifically for the ants, which is high in nutrients, in addition to the nectar the plant produces in its flowers for pollinators. In exchange for their food and shelter, the ants protect the plant from invaders like herbivores, vines, and other insects that could threaten their host plant by attacking it and eating it.

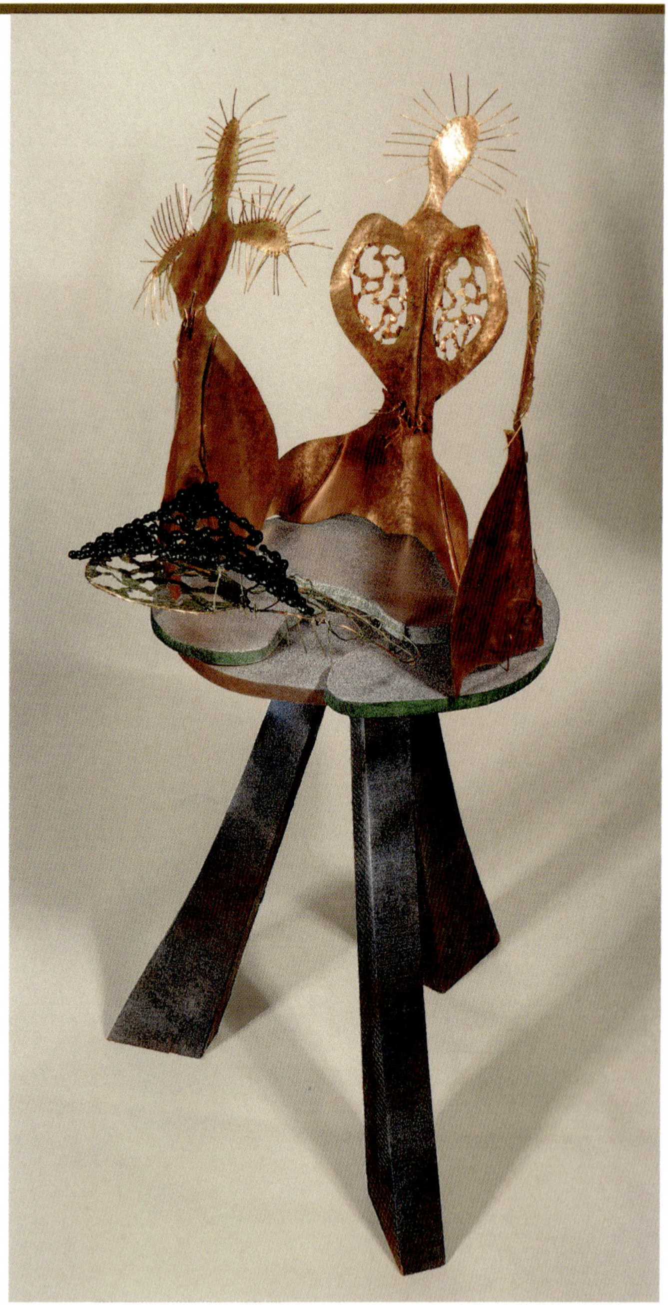

"Air and Ant" shows air plants reaching towards the sun with tendrils outstretched. The rough walls of the interior chambers are shown in the center of the sculpture, with the dark bodies of the ants swarming at the base to protect the plant. (copper)

Bark and Vine

Paper trees in the jungles of Ecuador are sometimes threatened by strangling vines. The paper tree protects itself by expanding and shedding its bark, which snaps the vine's stranglehold. The vine breaks and dies. This is just one example of the sophisticated adaptation that exists within the plant world to help competing species maintain a sustainable balance, ensuring their survival.

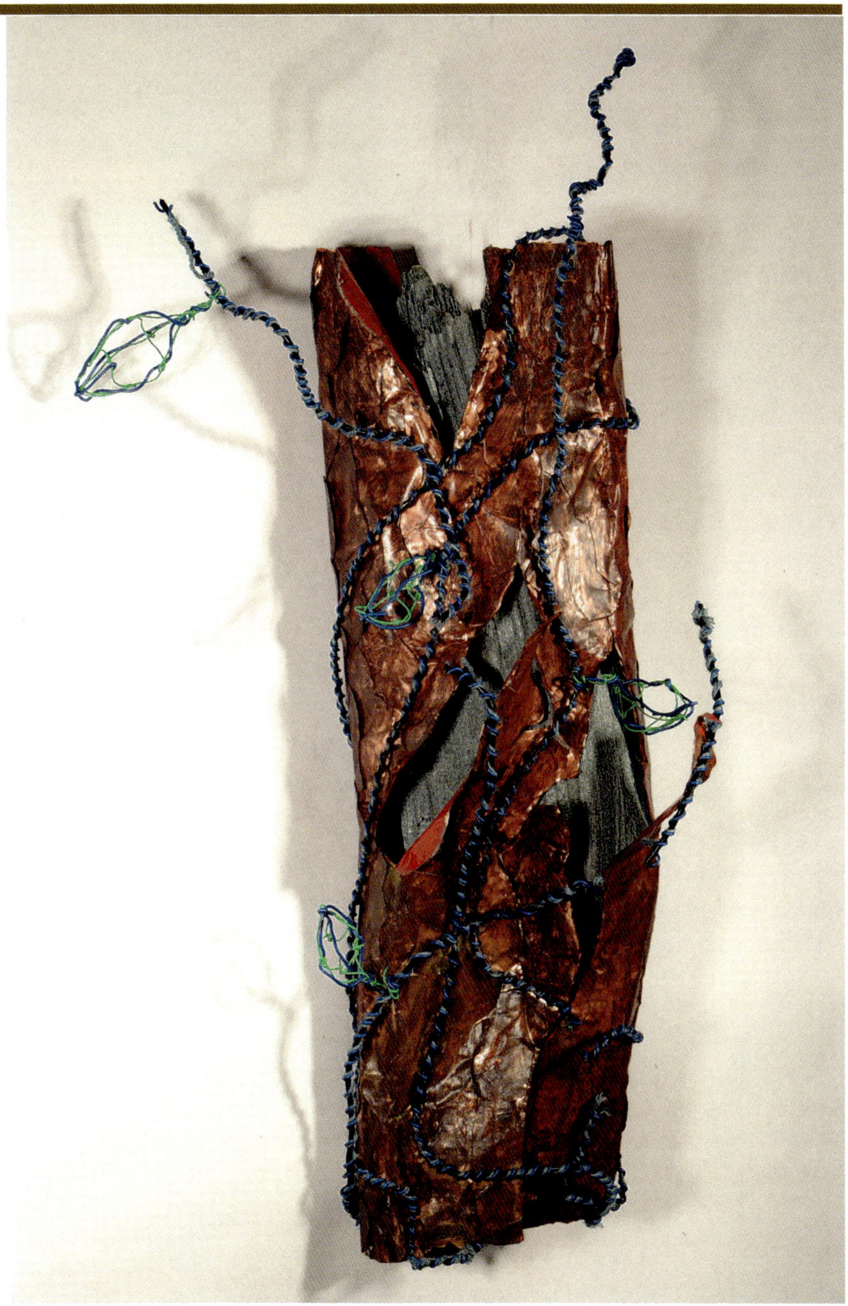

"Bark and Vine" shows the copper bark pulling away from the interior layer of the tree, leaving the vines broken and dying. The vines are constructed of thick wire with thinner wire spiraling around the core. (copper, wire)

Angel's Trumpet

Brugmancia, or Angel's Trumpet, is a hallucinogenic plant of the indigenous shamans of Ecuador, used in ceremonies and rituals. Parts of this plant are highly toxic and can be fatal if ingested. It is said that the plant earned the name "Angel's Trumpet" because the last sound a person would hear after ingesting the plant would be the sound of celestial trumpets. The "trumpet" of the plant is long and deep and it takes a very special pollinator to insure the plant's survival. The sword-billed hummingbird is the only bird that can pollinate this plant. It has developed an extra-long beak in order to feed on the nectar that is deep inside the plant. The deadly plant and the tiny bird have achieved a delicate balance.

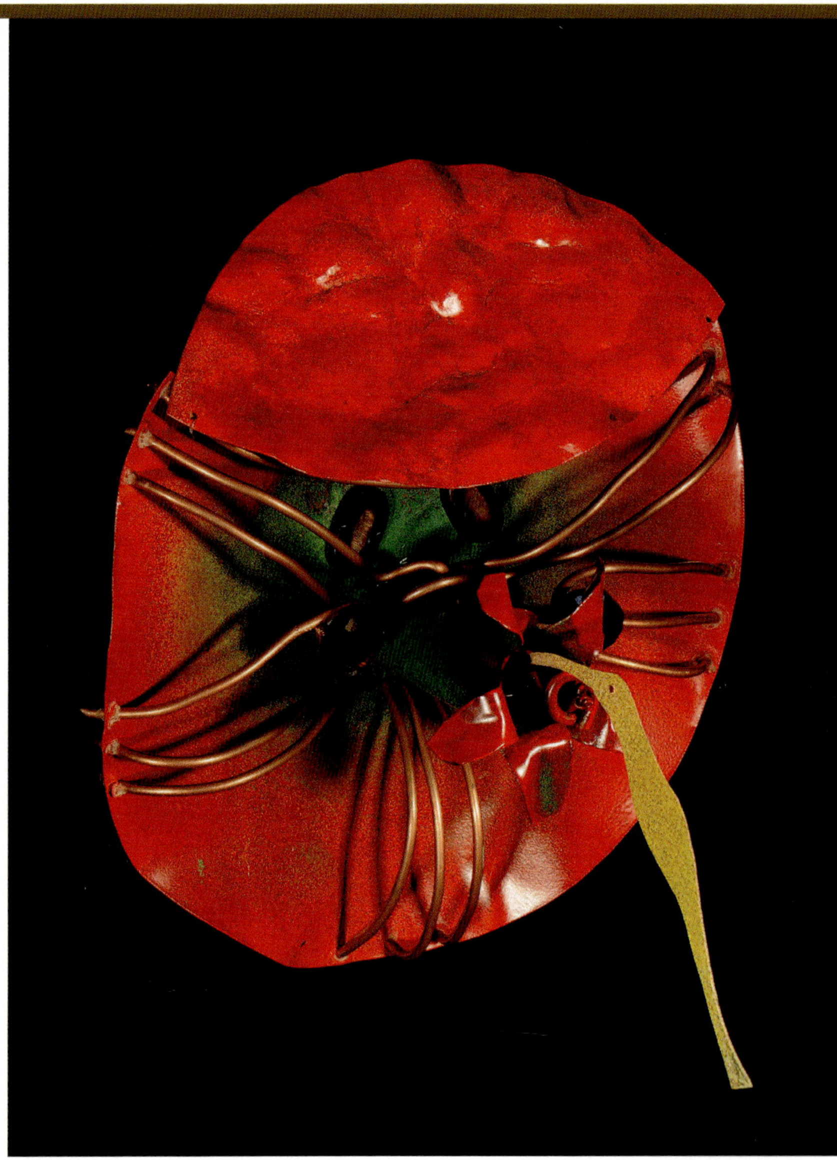

"Angel's Trumpet" shows the face of a shaman at the top of the piece. The copper wires represent the plant's graceful stamen. The green metal and wire shows the plant's petals, with the yellow hummingbird placing its beak inside the plant. (metal, wire)

Kudu and Acacia

The acacia is a drought-resistant tree on the African plains that is often the last source of food for the kudu antelope. When hungry herds flock to these trees and begin to decimate their leaves, the acacia have an amazing ability to protect themselves. First they release volatile airborne pheromone molecules that warn the surrounding acacia trees that a threat is present. This signals the acacia trees to increase their tannin production and fill their leaves with the substance. Tannin can be toxic if too much of it is ingested. The result is a culling of the herd, which reduces the antelopes' population to a sustainable size, allowing the trees to survive. The natural balance between the kudu and the acacia is restored.

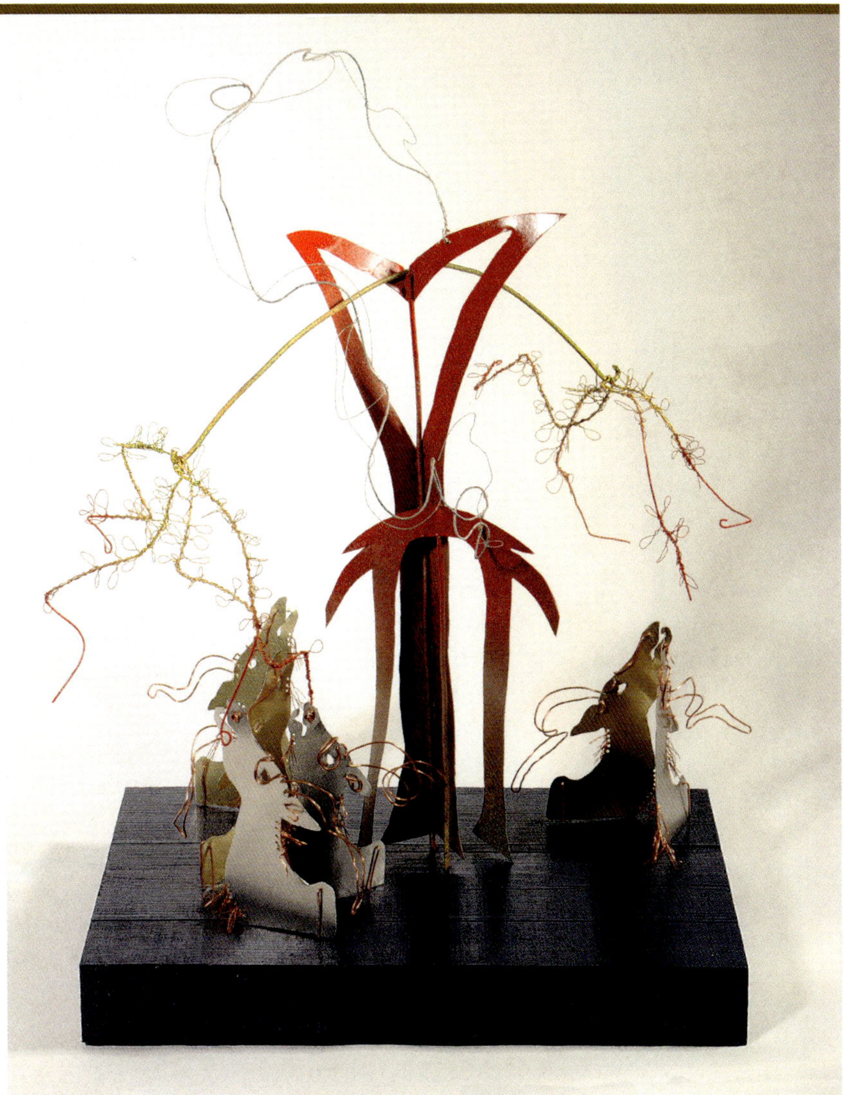

"Kudu and Acacia" shows an acacia tree surrounded by hungry kudu. The reaching twists of wire from the top of the tree show the acacia tree communicating the threat. The branches, stripped of their leaves, are above the outstretched necks of the kudu. The delicate wires at the top of the antelopes' heads represent the majestic curved horns of the kudu. (metal, wire)

Fir and Fungus

The mighty Douglas firs of the Pacific Northwest make up a vast dense forest. These enormous trees have found a sustainable balance with an underground network of fungus roots. The Douglas firs absorb CO2 and produce sugar. The trees are dependent on the mycelium root network of the fungus to take in the CO2 and sugar and process it into nutrients, which are then distributed to the neighboring fir trees. The mycelium are the fine white filaments of the fungus root system, and they form an enormous network underground, connecting the stand of fir trees. This connective web is considered one of the biggest living organisms on the planet.

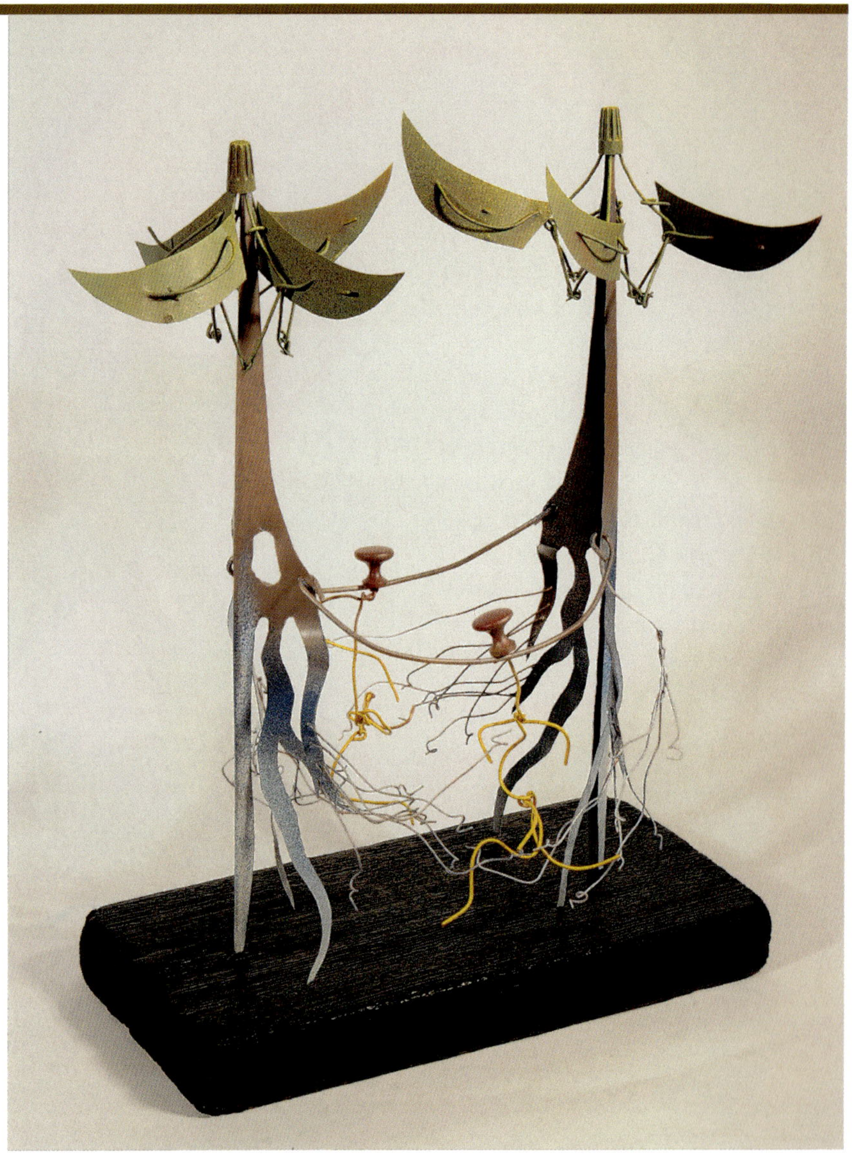

"Fir and Fungus" shows Douglas fir trees standing separately and towering in the sky. The metal knobs on this sculpture represent fungus, and the fine wires connecting the two trees represent the connective mycelium web that brings nutrients to each of the fir trees. (metal, wire)

Grouper and the Moray Eel

The grouper fish and the moray eel have a balanced, cooperative relationship when it comes to hunting for food. The grouper is very quick at catching prey in the open water but is unable to maneuver through the rocky bottom crevices, coral, and caves that are home to the moray eel. The eel is very effective in catching prey in and amongst the rocky terrain. The two species have learned to combine their hunting efforts in a way that benefits them both. When the grouper spots a fish or school of fish amongst the rocks, it starts to shimmy its body. The shimmy is followed by the grouper standing on its head and tossing its head back and forth, pointing in the direction of the prey fish. That gesture gets the attention of the moray eel, which looks in the direction the grouper is "pointing." The eel takes charge and drives the prey fish out into the open water where the grouper can grab them. The fish are then shared by the grouper and eel.

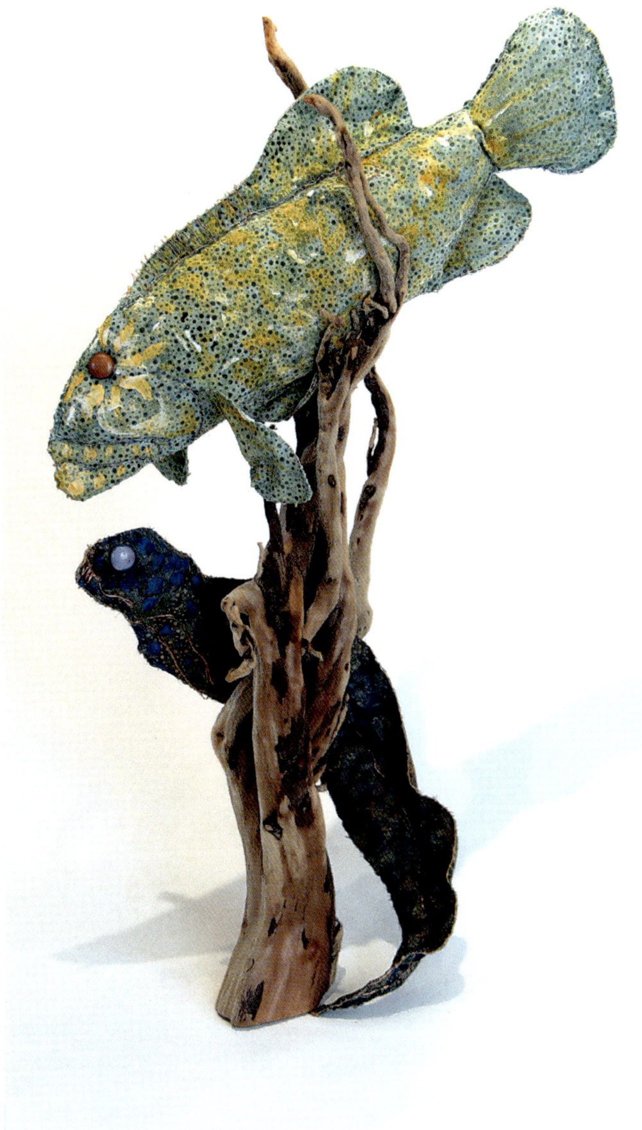

"Grouper and the Moray Eel" shows a grouper, made from fabric, swimming through a coral reef, made from driftwood. The moray eel, also made from fabric, swims below, waiting for a signal from the grouper that dinner is close by. (fabric, driftwood)

Leafy Seadragon

The leafy seadragon lives a balanced peaceful life, floating amongst the seaweed off the coast of Australia, carried along by the currents. It is perfectly camouflaged, with the leafy projections from its body matching the undulating shape of the seaweed. It is related to the pipefish and seahorse, but unlike them cannot use its tail for holding on to marine plants. There are some similarities with seahorses, however. The male seadragon plays a major role in carrying and hatching young seadragons. Instead of depositing her eggs in a pouch of the male, like the seahorses do, the female seadragon lays her eggs under the male's tail on a spongy brood patch. As the eggs are laid the male fertilizes them, then carries the eggs to term, which takes between 4 to 6 weeks. Finally the tiny seadragons hatch and float into the safety of the seaweed. This creature is rare because its population is threatened by the pet trade.

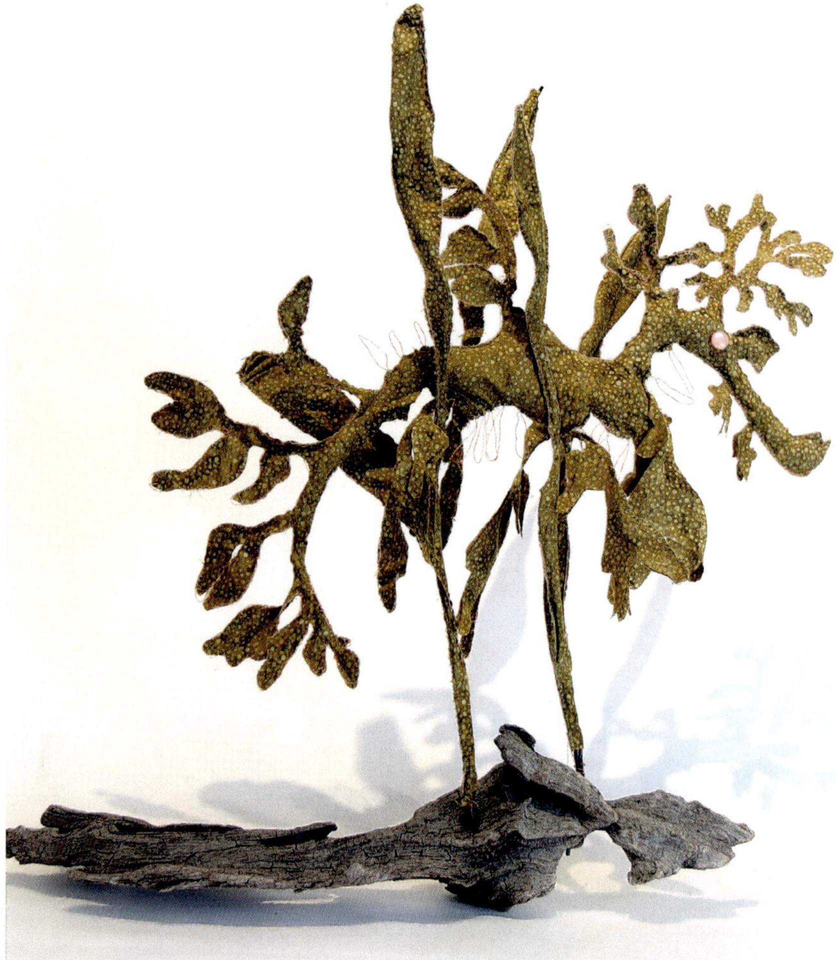

"Leafy Seadragon" shows a leafy mass of seaweed with a leafy seadragon nestled amongst it. Look to the right side of the sculpture to see the seadragon's eye. The floor of the ocean is represented by a piece of driftwood. (fabric, driftwood, wire)

The Satanic Leaf-tailed Gecko

The satanic leaf-tailed gecko has perfected the art of camouflage to live in balance and harmony in the rainforests of Madagascar. Its skin assumes the color and texture of leaves to hide from predators, and when viewed from above, its tail's shape is almost indistinguishable from the surrounding leaves of its hiding place. It is now a threatened species because of its popularity in the pet trade, driven by its remarkable ability to look like a leaf. The gecko is nocturnal, and in addition to camouflage, it has several other techniques to insure its survival: The gecko has the ability to flatten its body against the leaf litter, looking even more like a two-dimensional leaf. If a predator approaches, the gecko opens its mouth to expose a bright red interior to frighten off the predator. (This is probably the reason behind naming this small reptile "Satanic.") What makes this creature unique is its ability to be camouflaged amongst the leaves, both by its coloration and the shape of its remarkable tail. This magical skill ensures its survival.

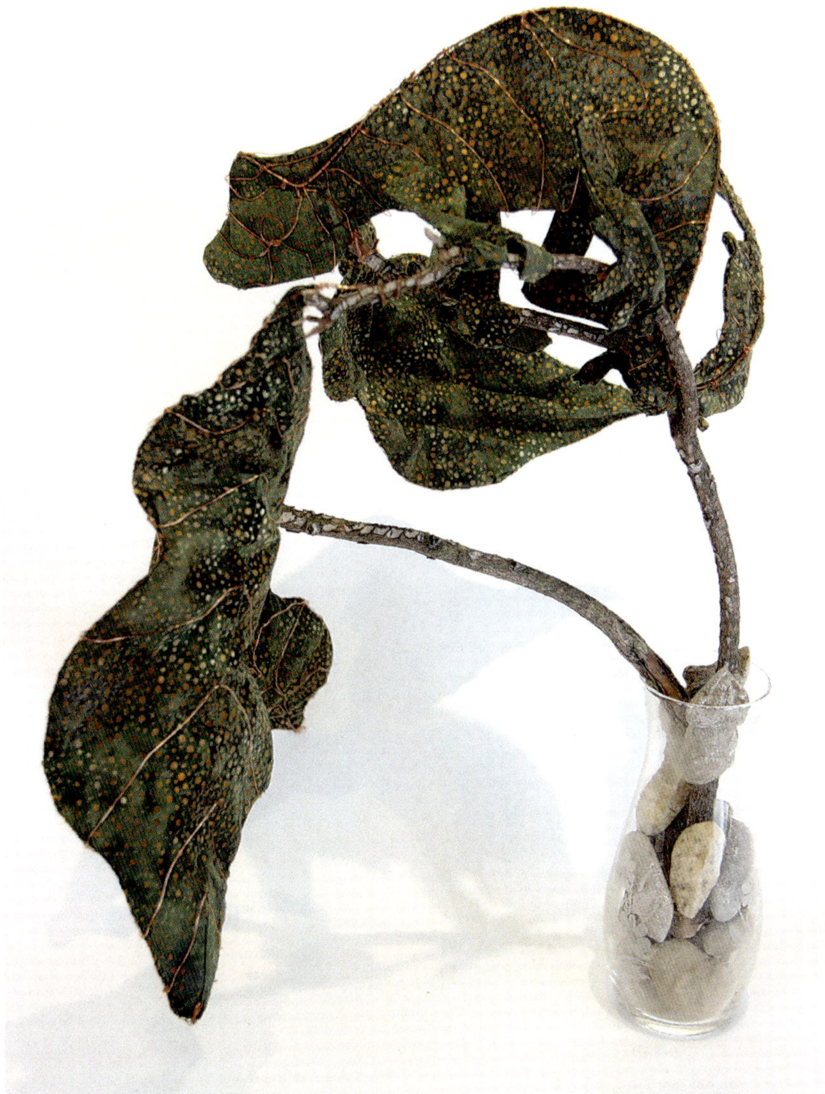

"The Satanic Leaf-tailed Gecko" shows brilliantly colored fabric leaves on wooden stems. The satanic leaf-tailed gecko is at the top of the sculpture, expertly camouflaged from its predators. (fabric, wood, wire)

Kin Recognition

The Arabidopsis thaliana plant is a small flowering plant, native to Eurasia. Considered a weed, this plant has developed a remarkable survival technique: The plant has the ability to recognize its kin through a chemical secretion given off by its roots. When an Arabidopsis thaliana recognizes its kin growing next to it, the plant shortens and coils its roots to avoid competing with its neighbor. This allows both plants room to grow. If the plant does not detect the same species next to it, it grows its roots to their maximum length, competing aggressively for space and nutrients, achieving maximum growth. These amazing plants find balance in recognizing kin and cooperating to insure their mutual survival.

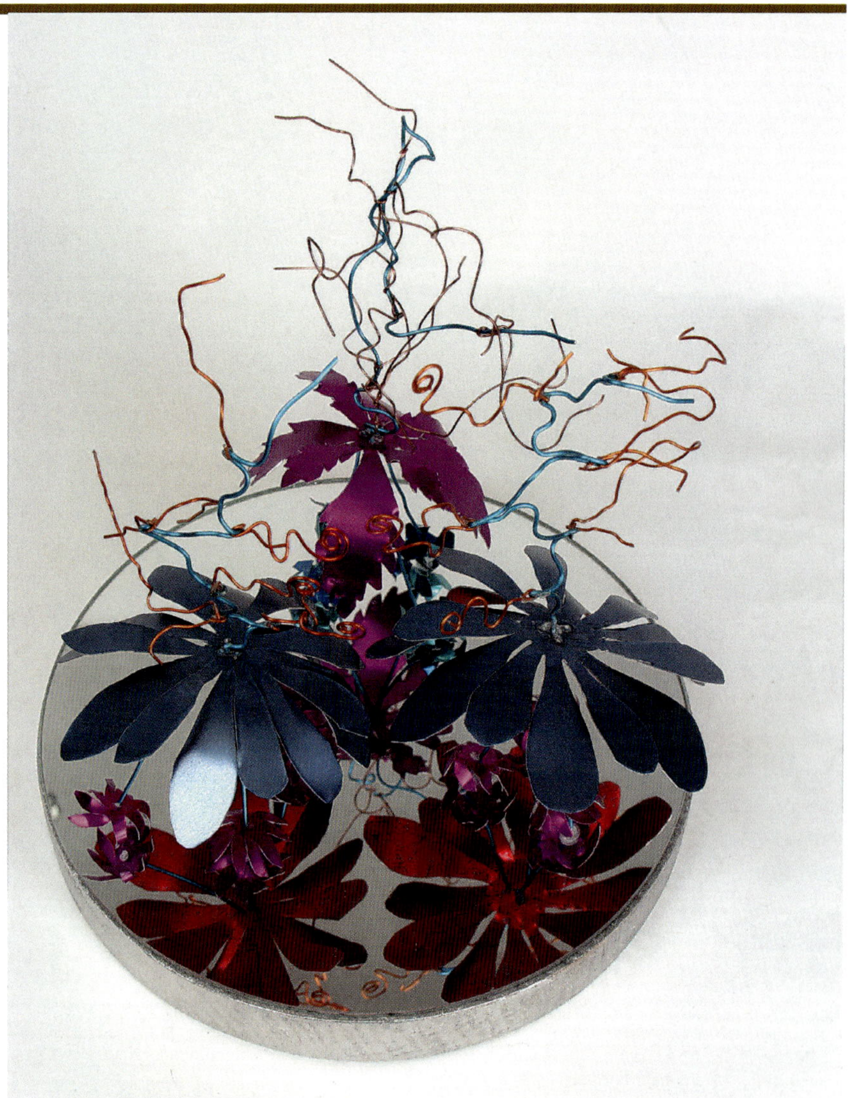

"Kin Recognition" is presented upside down to highlight the behavior of the plants' roots. The flowers and leaves rest on a mirror at the bottom of the sculpture. The plants that are kin have their roots curled facing each other, avoiding competition. The plant that is not the same species has its roots stretching out, competing for space and nutrients. (metal, wire, glass)

CLIMATE CHANGE

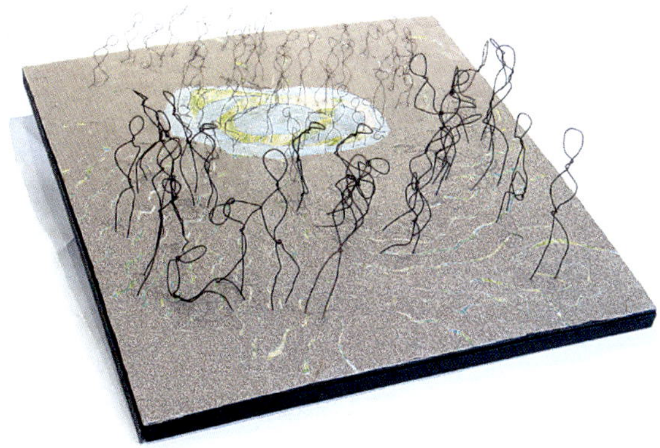

The Earth knows how to regulate its temperature. Evaporation, convection, and thermal radiation all act in concert to maintain the balance of heat coming from the sun and leaving the Earth. The greenhouse gases, notably CO2, have disrupted that balance. They absorb thermal energy and radiate it in all directions, including back towards Earth.

The amount of CO2 in the atmosphere fluctuated from 200 parts per million (ppm) to 300 ppm and back again during ice ages. This sustainable fluctuation changed after the Industrial Revolution began in 1820. The burning of coal and other fossil fuels increased CO2 in the atmosphere until it now measures over 400 ppm. CO2 has raised the temperature of the planet by more than 1 degree Celsius. The increased CO2 in the atmosphere is also changing the chemistry of the oceans, making them more acidic. Shell development of all juvenile shellfish is being negatively affected, and the coral reefs are also disintegrating. The habitat of fish, shellfish, and the protection that the reefs afford land from the ocean's power, is threatened.

Ice on the land and sea is melting, as are the glaciers. The disappearance of the white reflective ice has created a "feedback loop," resulting in additional warming of the land and sea because of the absorption of heat into the dark surface of the water and land. It is expected that the Arctic summer ice will be completely gone in a few years. The ice shelves in Antarctica and Greenland that currently hold back the glaciers are melting and breaking off, raising the prospect of the glaciers flowing into the sea and raising the water level worldwide by about 20 feet, affecting human settlements and coastlines. In addition to rising sea levels, the melting ice is adding enormous amounts of fresh water into the sea, changing the salinity and temperature of the water.

Climate change is responsible for increasingly extreme weather patterns, including flooding, drought, insect infestations, and a large increase of wildfires. It is wreaking havoc with Nature's balance and has contributed to the extinction of a huge percentage of the Earth's species. The possible solutions to address climate change, and the time, money, and effort needed to effect those changes, are the subject of debate. The vast majority of scientists agree that climate change is real and that the debate must be driven by facts. Science, supported by legislation, offers the best chance of reversing the effects of climate change.

Climate Change

Climate change is real, and yet the fossil fuel industry is funding a major denial effort through certain news media outlets and powerful political and business personnel. Their goal is to discredit the 97% of all climate scientists worldwide that are raising the alarm about this disastrous event. CO2, now measured at over 400 ppm, and other greenhouse gases in the atmosphere radiate the sun's solar energy causing the Earth's temperature to rise. Sea ice, polar ice sheets, and glaciers are melting, causing sea levels to rise. The melting of the permafrost releases tons of methane, the most dangerous greenhouse gas, into the mix. Climate change is already wreaking havoc, as witnessed by worldwide drought in the continental centers, intense wildfires, extreme weather, insect infestations, excessive heat, and the changing chemistry (acidification) of the oceans.

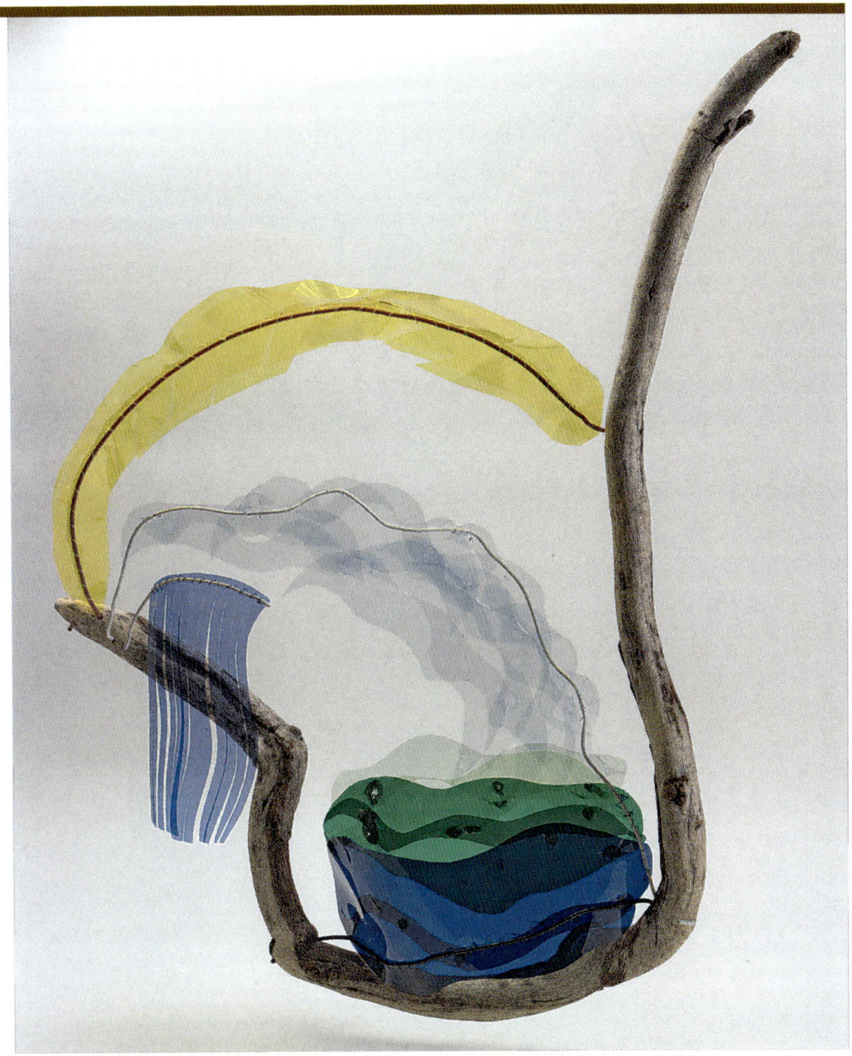

"Climate Change" illustrates the causes and effects of climate change. The yellow gel represents the dome of greenhouse gases that hold in the sun's heat. The land and sea rest in the base of the sculpture, with the heated ocean water evaporating and sending vapor rising into the air, as shown by the light blue gel. What goes up must come down, and depending on the time of year extreme amounts of either rain or snow falls back to the Earth, as depicted by the darker blue gel on the side of the sculpture. (theater gel, wire, driftwood)

CO2

CO2 is a major heat-trapping gas, contributing 60% of the emissions released into our atmosphere. The industrialized nations of the world create about 10 billion tons of CO2 annually. The oceans absorb two billion tons, plants absorb three billion tons, and the remaining five billion tons stay in the atmosphere, creating the problem of climate change. Ever since the Industrial Revolution began in 1820, fossil fuel industries have sent 240 gigatons of CO2 into the atmosphere, causing a 1.3 degree Celsius increase in the planet's temperature. Before 1820, there was an average of 275 ppm of carbon dioxide in the atmosphere. Now with the greater use of fossil fuels, for the first time ever we have increased the CO2 in the atmosphere to over 400 ppm. Fluctuations in CO2 used to occur in 100,000-year cycles during ice ages, as evidenced by the ice cores that climate scientists have extracted. Now these fluctuations are occurring in the span of only 200 years! The natural balance and cycles of all life on the planet, including human life, is being profoundly threatened by this rapid increase.

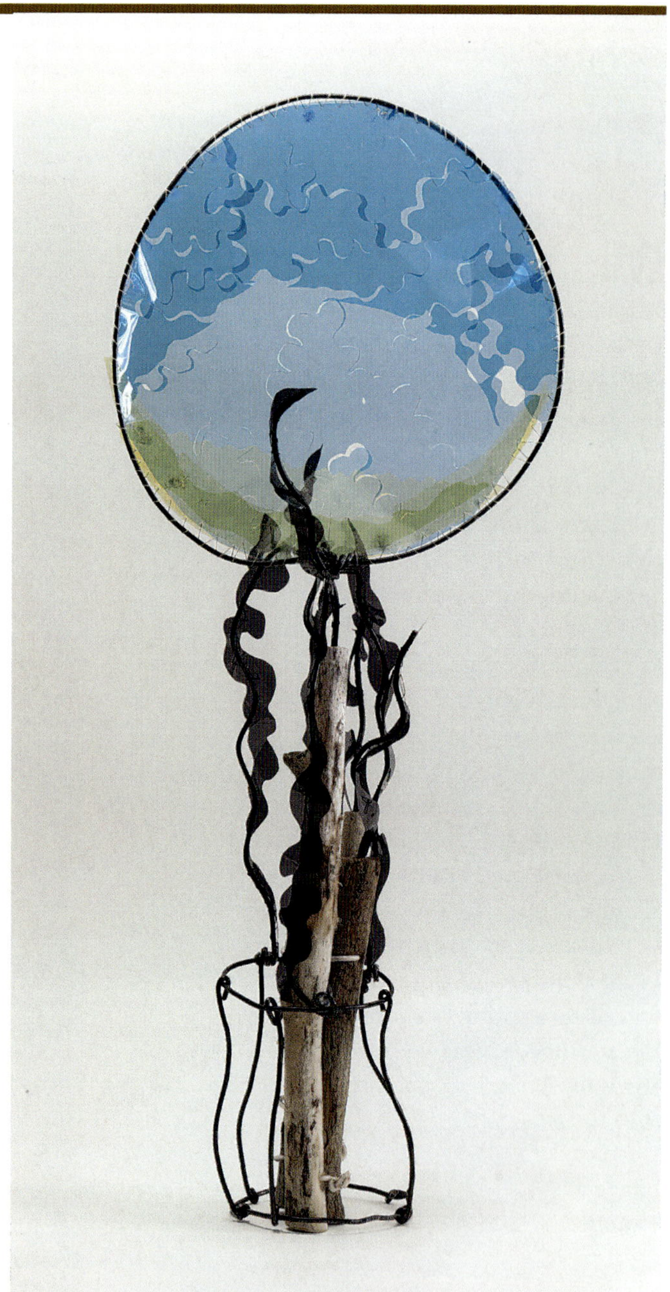

"CO2" shows the impact of CO2 on the Earth. Within the circle, the dark blue gel represents the enormous amount of CO2 in the atmosphere. The dark grey gels, coiling up from the base, represent carbon emissions from a power plant. (wire, driftwood, theater gel)

Methane Gas

Methane hydrate gas is found stored in relatively shallow deposits (less than 2,000 meters) in polar regions and under the ocean floor. It is a powerful greenhouse gas approximately 30 times more lethal per molecule than CO_2. Its life span is 12 years, compared to CO_2's life span of 20-200 years, but it is 105 times more effective than CO_2 in heating up the planet over a 20-year span. The stored amount of methane at the poles and under the ocean is equivalent to between 1,000-10,000 gigatons of CO_2. If this amount is released into the atmosphere, it has the potential to raise the temperature of the planet by as much as 4 degrees Celsius. Unfortunately, the enormous increase in fossil fuel use is warming the Earth, the oceans, and the polar ice. As the land and water warm, the permafrost and ice will melt, and the long-held methane will escape into the atmosphere. The release of this powerful greenhouse gas will exacerbate the effects of climate change.

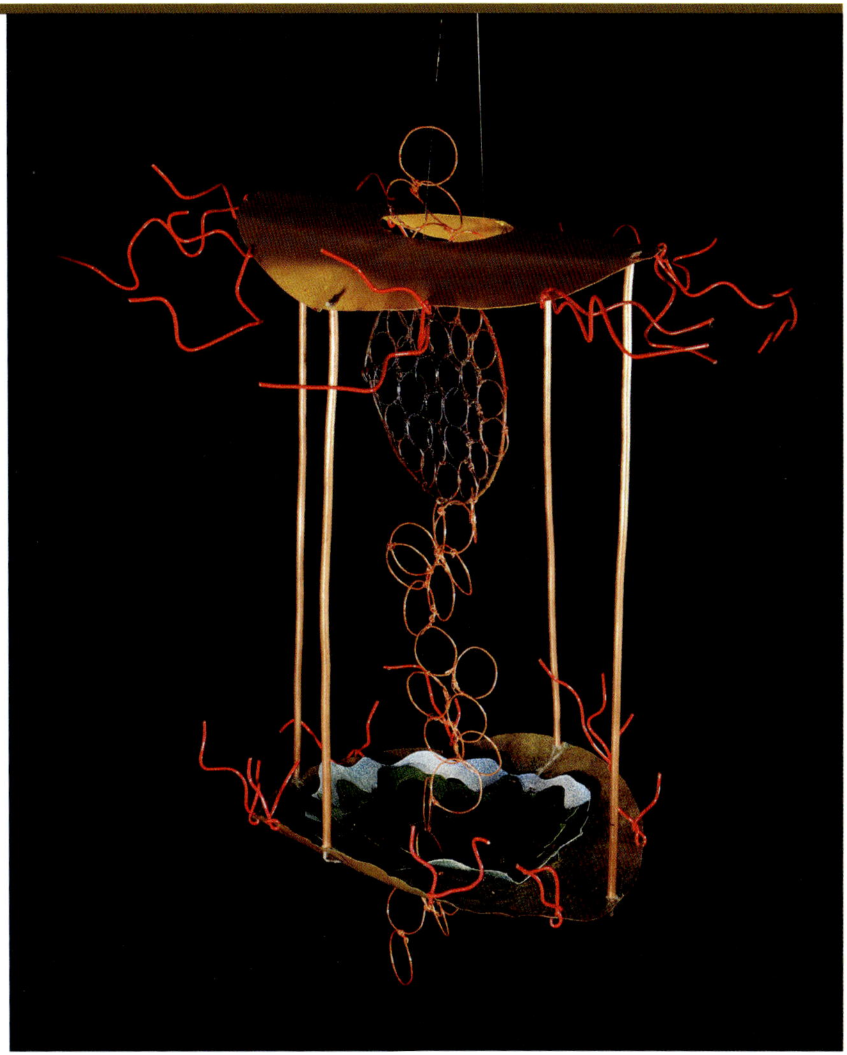

"Methane Gas" features a methane molecule, represented by the large circle with thirty smaller circles inside, showing how much more lethal it is than CO_2. The painted metal at the bottom represents the permafrost at the poles and the ocean floor. The circular piece at the top of the sculpture is the heated atmosphere above the Earth, holding the methane in, rather than releasing it to space. The red wires represent fire and heat caused by the trapped methane. (wire, metal)

Ice Melt

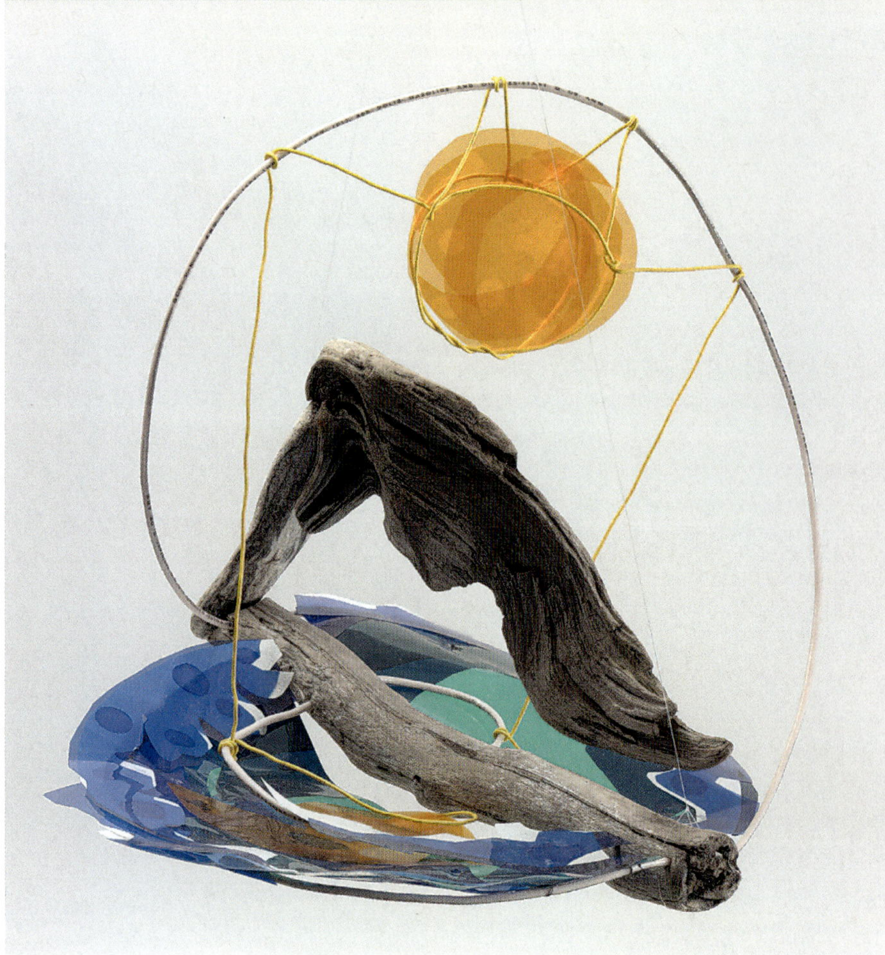

Record-breaking heat over the last several years has impacted the polar regions even more than the land. And unlike the white ice that reflects the sun's heat back out to space, the dark Arctic Ocean absorbs the sun's heat, raising the water temperature. In addition to the warmer water, excess CO2 in the atmosphere is being absorbed into the oceans, changing the chemistry of the water and making it more acidic. Polar animals are affected by the disappearing ice and the increased acidity in the water. As the ice floes move farther and farther offshore, polar bears find they cannot swim out that far, and as a result can no longer hunt for seals. Walruses also depend on the ice for fishing and for caring for their young. Pteropods, tiny sea snails, are being affected by the changing acidic chemistry of the oceans. Their shell development is weakened to the point of dissolving. As these tiny creatures are close to the bottom of the food chain, their diminished population will be felt all the way up the food chain.

"Ice Melt" shows the sun beating down on the Earth and the ocean, raising the temperature to an unsustainable level. Blue theater gel represents the melting sea ice pouring into the oceans. The arching driftwood represents the arctic winds, also a contributing source. (theater gel, string, wire, driftwood)

Acidification

The world's oceans are becoming more acidic, with their pH changing due to climate change. The oceans are like sponges, absorbing CO2 from the atmosphere, and dramatically changing their chemistry. The rising acidity impacts pteropods and other shelled creatures, including krill. Their shells are weakened by the acidity, threatening their ability to survive to adulthood. As these creatures are near the bottom of the food chain, and are prey to creatures as diverse as whales and many species of fish, their loss will significantly affect the ocean's animal life. The increased acid in the water is also damaging coral reefs worldwide, weakening their structural integrity. These reefs protect a huge number of landmasses from the harsh onslaught of the ocean's force, and are the nursery habitats for many forms of sea life. The oceans cover 71% of the Earth's surface, and the impact of their changed chemistry due to the increased acidity is of enormous concern to scientists.

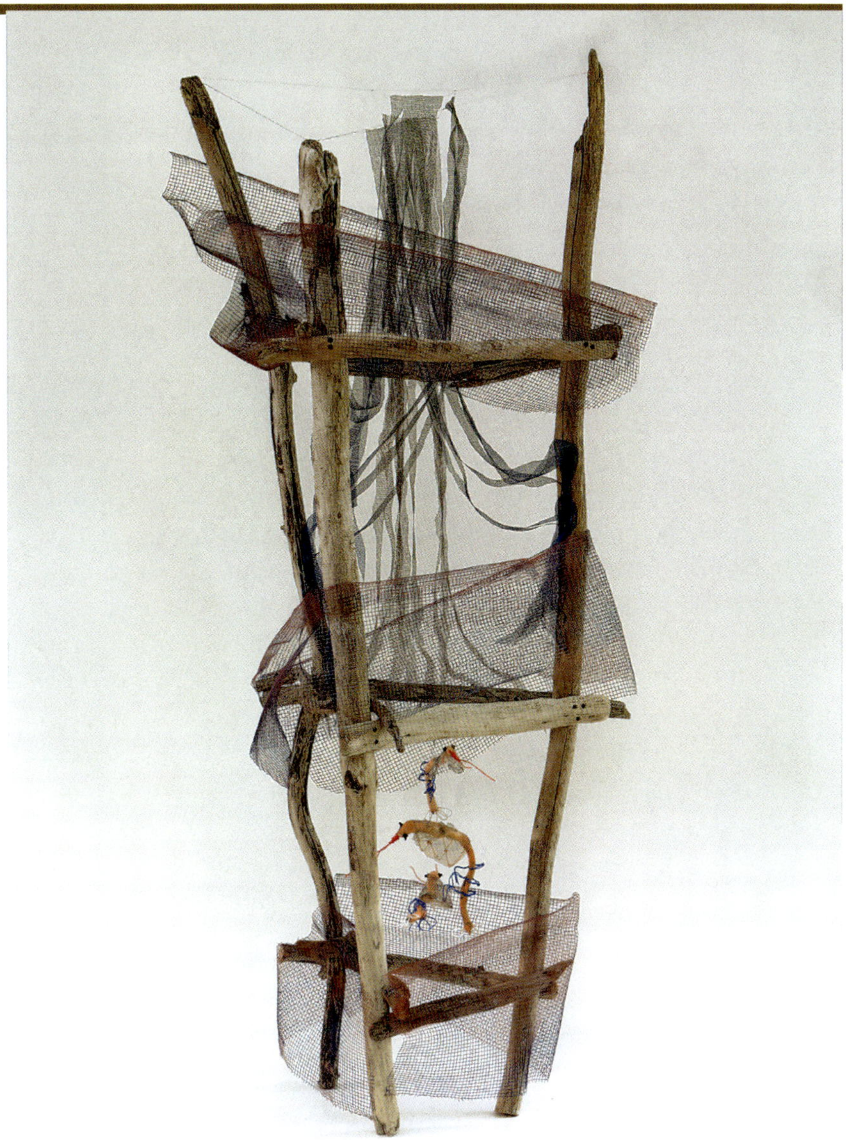

"Acidification" features three krill, which are close to the bottom of the food chain, shown in various stages of disintegration due to the increased acidity in the ocean. The vertical sweeping blue-gray screening is the added CO2. (wire, driftwood, theater gel)

Hunger Strike

The spruce bark beetle is only the size of a grain of rice, but it is capable of causing devastating damage to its host trees. The spruce beetle is found in the boreal forests of the northern hemisphere. Its numbers are balanced and sustainable in average temperatures; however, when the temperature rises, the insect experiences a population explosion. Climate change has raised the temperature in the north enough to cause an infestation of the spruce bark beetle. The result is that millions of acres of spruce trees have been blackened and decimated by this tiny beetle.

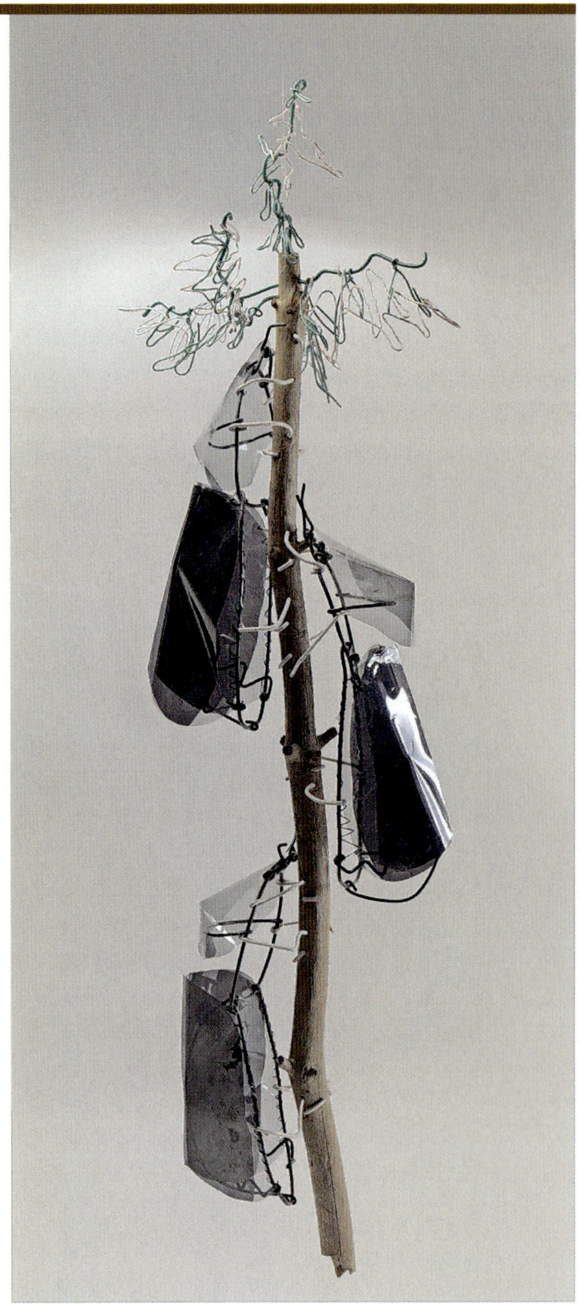

"Hunger Strike" shows three spruce bark beetles climbing to the top of a spruce tree. The beetles are shown disproportionately large to show the population explosion of this species. The "tree trunk" stick is bare except for the top's few remaining branches, representing the almost complete destruction by these hungry beetles. All of the beetles are focused on those branches, climbing towards the last living part of the spruce tree. (wood, theater gel, wire)

Baseball-Sized Hail

Climate change causes extreme weather patterns. In addition to violent rainstorms that bring flooding and high winds, these ferocious storms can bring tornadoes and hail. Mile-wide tornadoes and hail the size of baseballs were reported in Oklahoma and Texas recently. Meteorologists blame the extreme temperature change from the surface of the Earth to the atmosphere for creating the conditions that led to the unusually large hailstones. The hail in Texas pummeled the area for two days, smashing car windshields and windows in homes, and destroying solar panels. Schools received major damage, forcing their closure until the roofs could be repaired, and the flood waters receded. No lives were lost in those incidents, but the damage to property was significant.

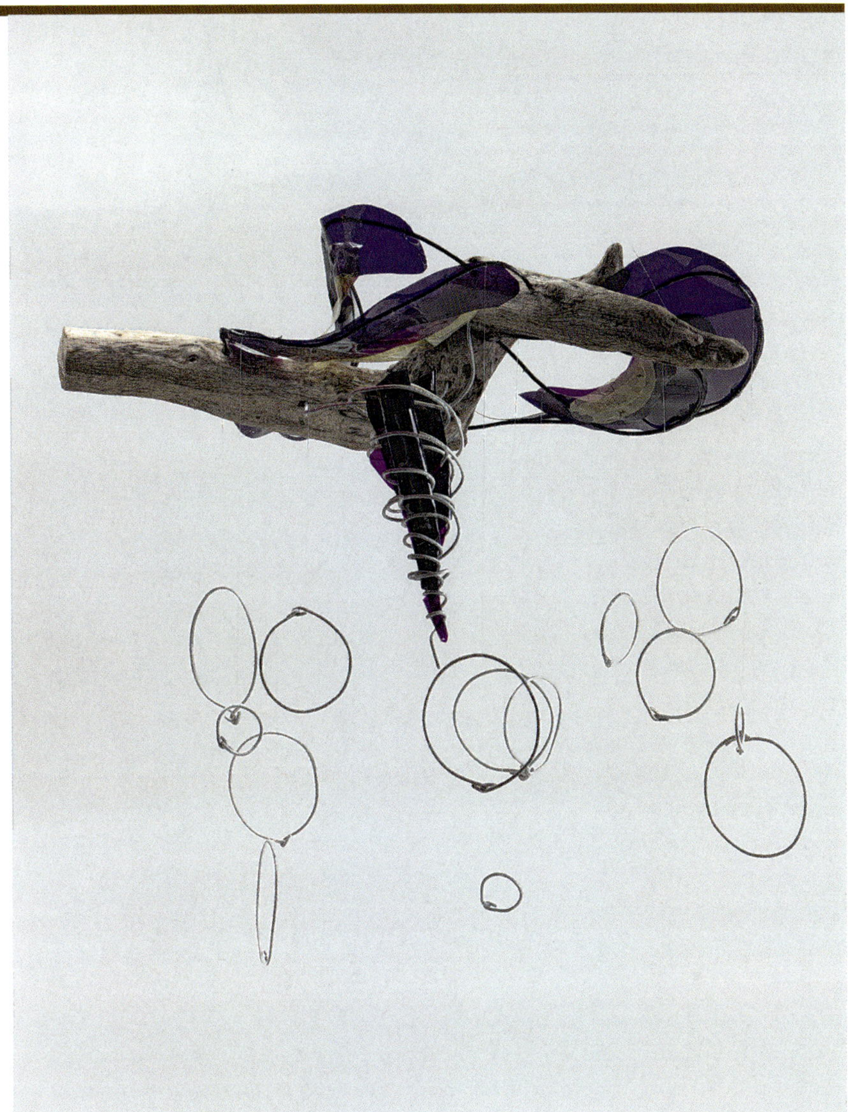

"Baseball-Sized Hail" illustrates the violent weather in Oklahoma that created the damaging hailstones. The largest of the round hanging wire circles are the actual size of baseballs, to demonstrate just how huge these hailstones actually were. The tornado in the center of the sculpture is made of dark purple gels and swirling wire. (wood, wire, theater gel)

High Water

Climate change causes the sea level to rise in two ways: Ice on the land melts and flows into the sea, and the warming temperatures in the water cause the sea water to expand. The interior ice sheets and glacial ice of Greenland and the Antarctic are melting. If these glaciers completely slide into the ocean it has been predicted that the sea level will increase by at least 20 feet, and 70% of the Earth's coastlines will be impacted. Currently Miami, Florida, is dealing with tidal flooding on a daily basis.

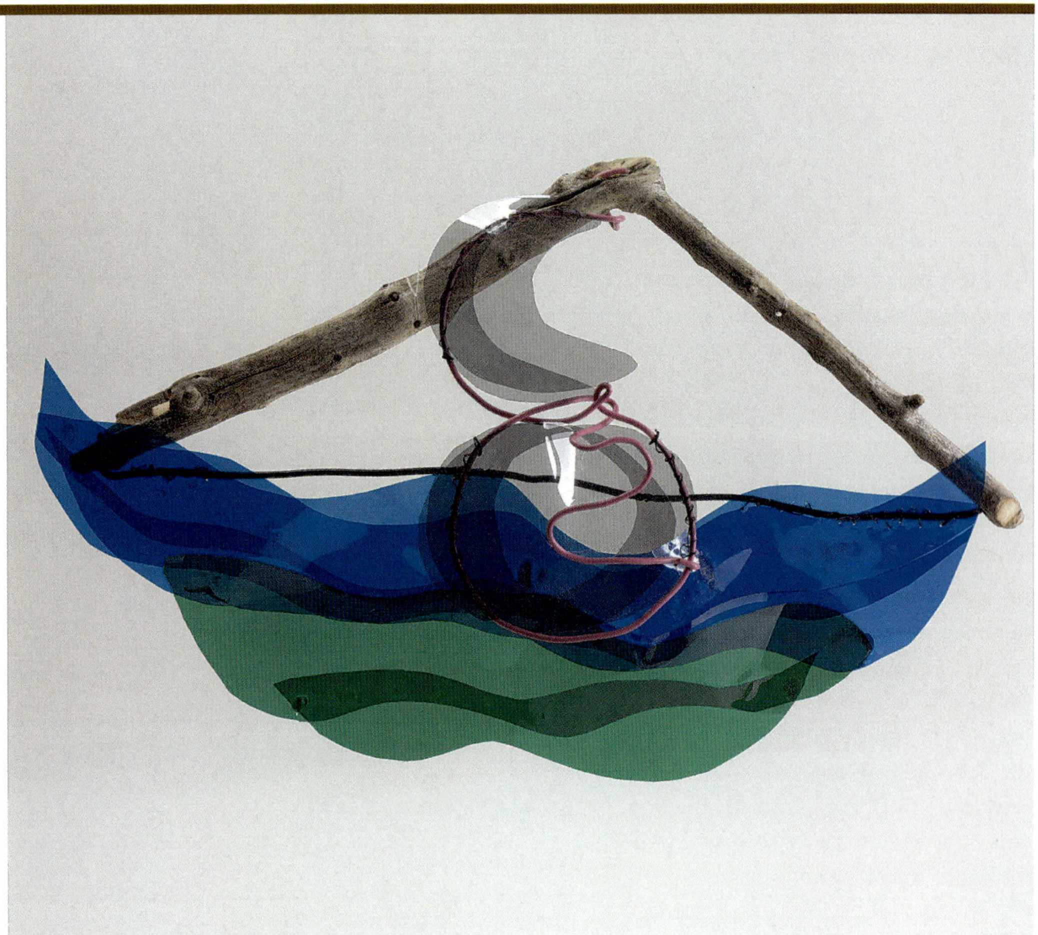

"High Water" is about the rising water associated with the worldwide great ice melt. The shape of the driftwood resembles a rooftop. The blue-gray theater gel represents CO2 floating outside the roof, and the green and blue gels represent the high water that is flooding the house. The catastrophic flood in New Orleans comes to mind, as do the threatened coastal cities of Florida. (driftwood, theater gel)

Heat & Fire

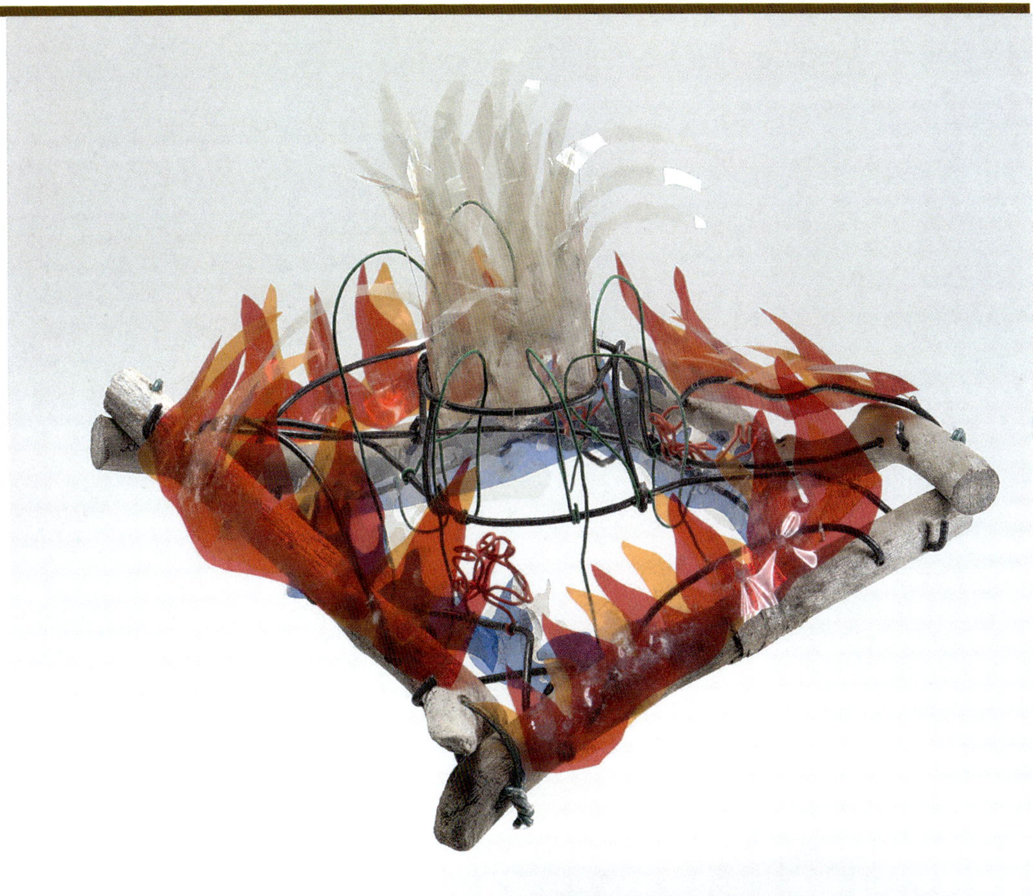

Mother Nature is turning up the temperature: All over the world cities are experiencing extreme heat and Moscow, Russia, is no exception. The record-breaking heat in 2010 in that city caused people to jump into the plaza fountains, fully clothed, desperate to cool down. The blistering temperatures fueled extremely dry conditions that fed out-of-control fires that burned neighboring Russian villages to the ground. Moscow was surrounded by fires, and the smoke drove people with respiratory conditions into their apartments to escape the thick polluted air. Unfortunately, many had no central cooling and so the heat inside was worse than outside, driving them back out into the extreme conditions. There was no safe place in Moscow that summer.

"Heat and Fire" shows a fountain with water spraying out of the top and raining down on three people in the bottom tier. Fire is boxing in the city, trapping the residents. (wire, wood, theater gel)

Migration

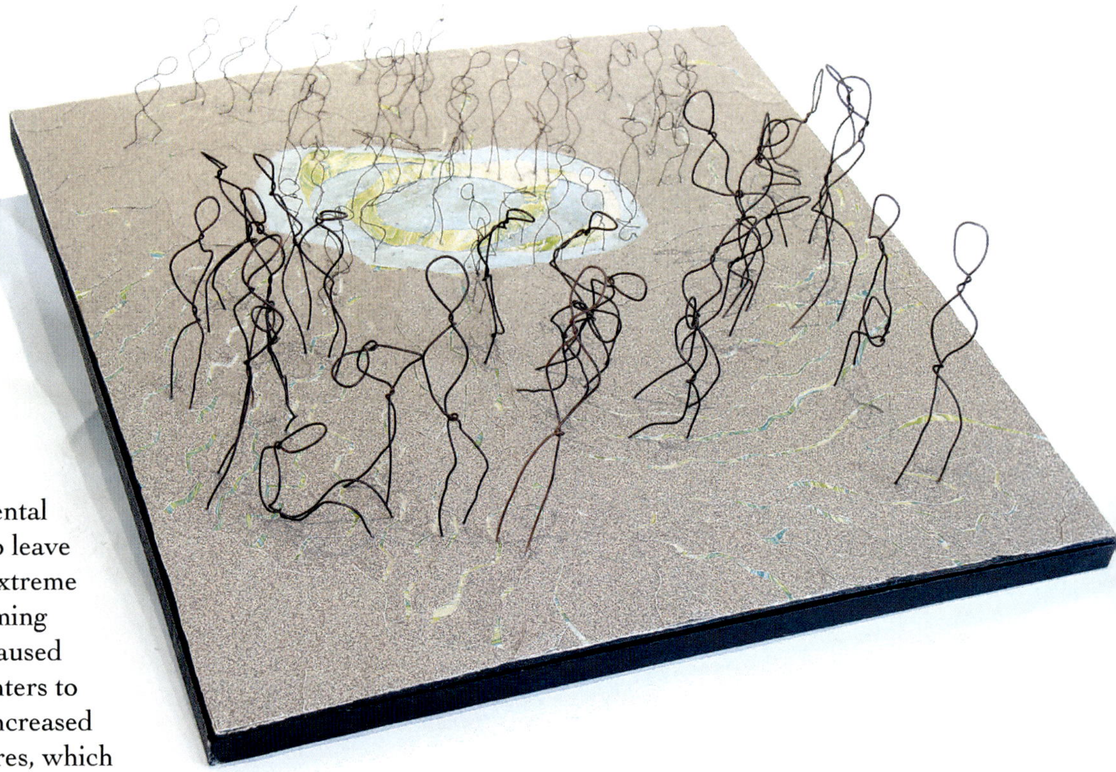

Climate change is affecting the land, water, animals, birds, and fish. It also affects humans. Millions of people worldwide have become environmental migrants, forced to leave their land due to extreme drought. The warming of the planet has caused the continental centers to dry out from the increased surface temperatures, which causes increased evaporation, lower precipitation, and reduced moisture in the soil. Farmers are unable to grow their crops. They have no product to sell, and no food to feed their families. These desperate families are leaving their homes in the hope of finding a sustainable existence elsewhere.

"Migration" is a stark depiction of thousands of people leaving their homes in search of land that will sustain them. The wire silhouettes of the migrants show some staggering, and all appearing lost in the parched barren world that was once their home. (wire)

ENVIRONMENTAL DEGRADATION

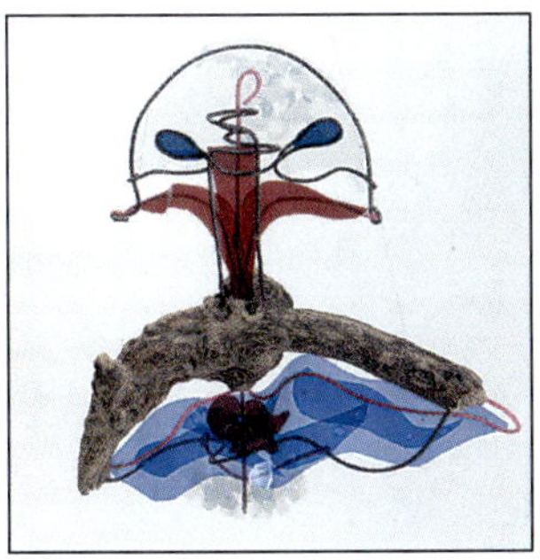

Man's impact on the Earth has steadily degraded the environment for humans, plants, and animals, to the point where the sustainability of the Earth as we know it is threatened.

This is the largest section of the book, because environmental degradation is occurring everywhere we look: above and below the ground, in the oceans, in the air we breathe and the water we drink, and in our own backyards. The art pieces on the following pages speak to the multiple reasons that so many species are threatened. Loss of habitat, pollution, poaching, and the increase of heat from climate change are all having a huge impact on Nature. Pollinators are dying and so our crops are affected. Sixty-seven percent of all the species on Earth have become extinct in just the last 20 years, an extinction rate and speed never before witnessed on Earth. There are now over seven billion human beings on the planet, so changing our way of life to protect the Earth will be a massive undertaking.

Crops

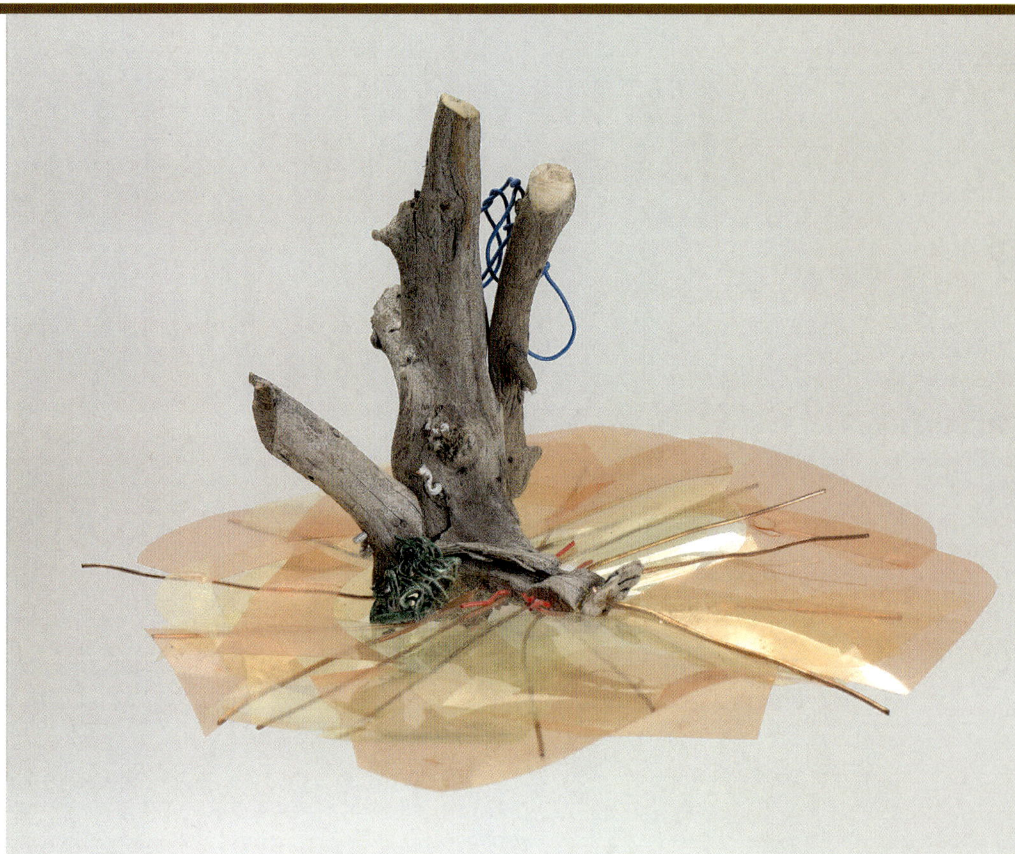

Deforestation of the world's rainforests is happening at an alarming rate. It is estimated that 80,000 acres of rainforest are cut down daily to provide open land for growing crops. The deforestation has impacted the flora and fauna of the planet. More than 135 plant, animal, and insect species perish daily, which equates to 50,000 species a year. In addition to the loss of habitat, the CO2 that is usually absorbed by the trees is now released entirely into the atmosphere and oceans. A recent photograph showed the stark reality of the loss of rainforest to farmland: The photo was in color and showed empty reddish plots of tilled farmland which filled the photo almost entirely, except for a tiny clump of green trees in the very center. After close observation it became clear that these few trees were all that was left of the rainforest.

"Crops" depicts the cutting of the forest and destruction of the land for growing crops. The small amount of green next to the cut stump represents how little forest is left. The ochre gels are the barren farmland and the blue wire tears speak for themselves. (wood, theater gel, wire)

Lung Tree

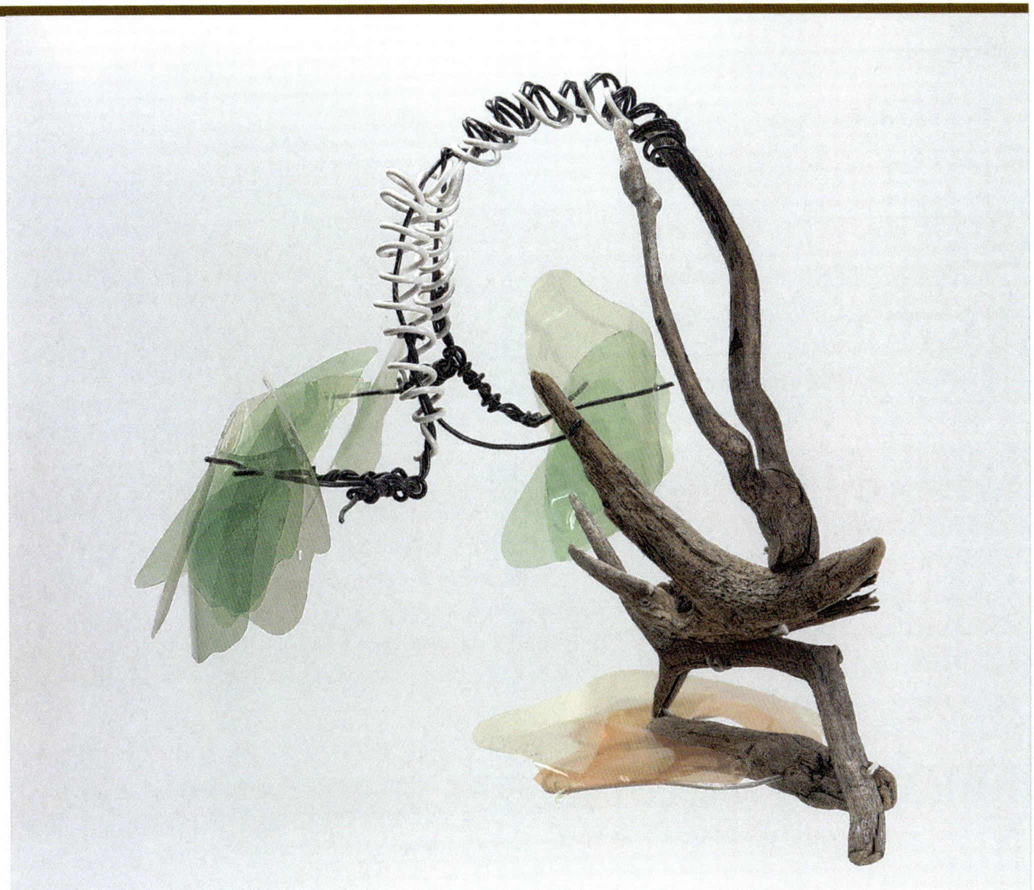

Prince Charles observed: "Forests are the world's air-conditioning system—the lungs of the planet—and we are on the verge of switching it off." The forests of the world are being decimated for man's use. The rainforests in particular are vital to the well-being of the planet. Trees absorb CO2 and give off oxygen, breathing in and out, in synchrony with how humans breathe. Humans and the planet need the oxygen that the trees, the "lungs," exhale. The rainforests affect the weather of the Earth by their absorption and creation of both rainfall and atmospheric gasses, and the daily loss of acreage in the rainforests is damaging the very air that the Earth breathes.

"Lung Tree" explores the notion that trees are the ecological lungs of the planet. The leaves are the shape of human lungs, creating a tree in the process of collapse, symbolic of the continuing destruction of the world's rainforests. (theater gel, driftwood, wire)

MONARCH BUTTERFLIES

Perhaps one of the most magical events in nature is the metamorphosis of a caterpillar into a butterfly. Butterflies, depending on their size and their species, usually live for a few days to a few weeks; however, the monarch butterfly hatches several generations each summer, living 3-6 weeks each. But the last generation of the summer lives for an astounding 8 months. These butterflies migrate from their northern birthplace south across the North American continent. Some of the monarchs overwinter in Mexico and some overwinter in southern California. Miraculously, these monarchs make the long migration north again in the spring, to their original host plant, and the cycle resumes. The spectacular clouds of orange butterflies filling the air and roosting in the southern trees is one of the greatest sights in nature. Unfortunately because of habitat loss, deforestation, excess heat from climate change, and the use of pesticides, this magnificent creature is dwindling in numbers from the billions that used to be to just millions today.

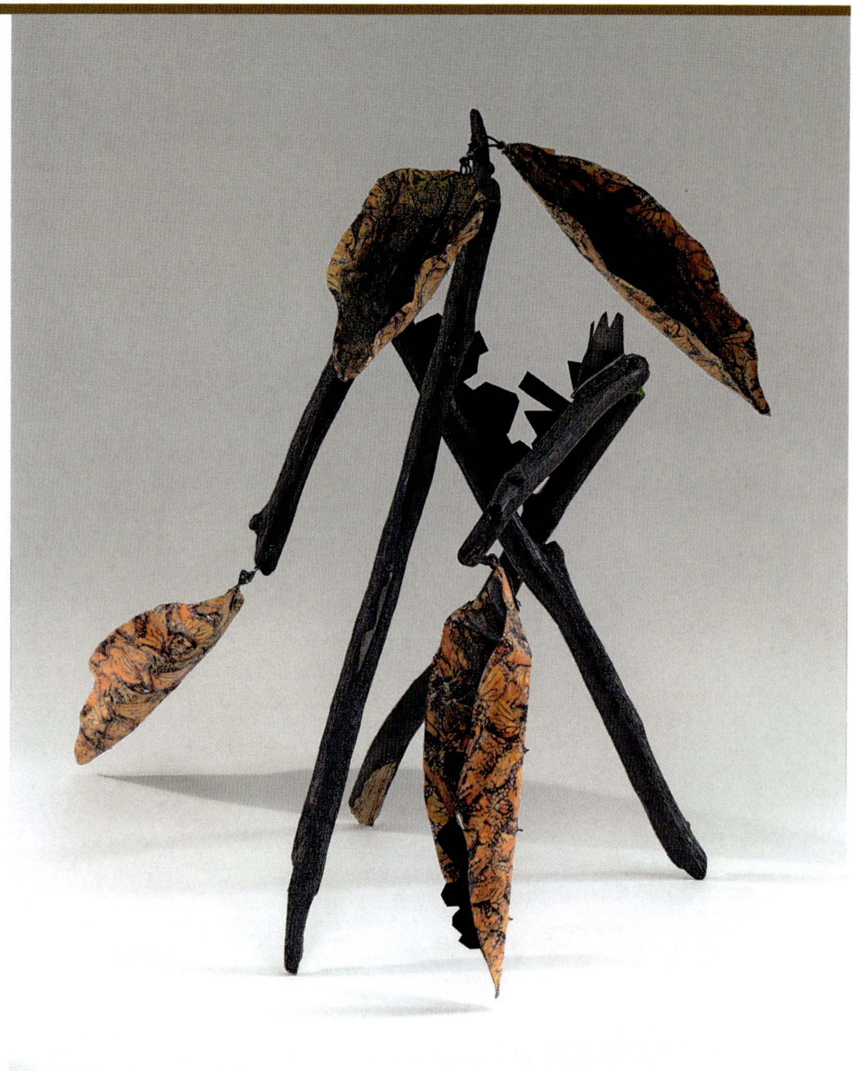

"Monarch Butterflies" shows the destruction of the monarch's habitat in exchange for housing development for human needs as shown in the black silhouettes placed in the pods. Monarch caterpillars depend on milkweed plants for food and protection. A collage of Monarch butterflies fill the milkweed pods as they prepare to lay their eggs for the next generation of this remarkable species. (metal, wire, wood)

Bee

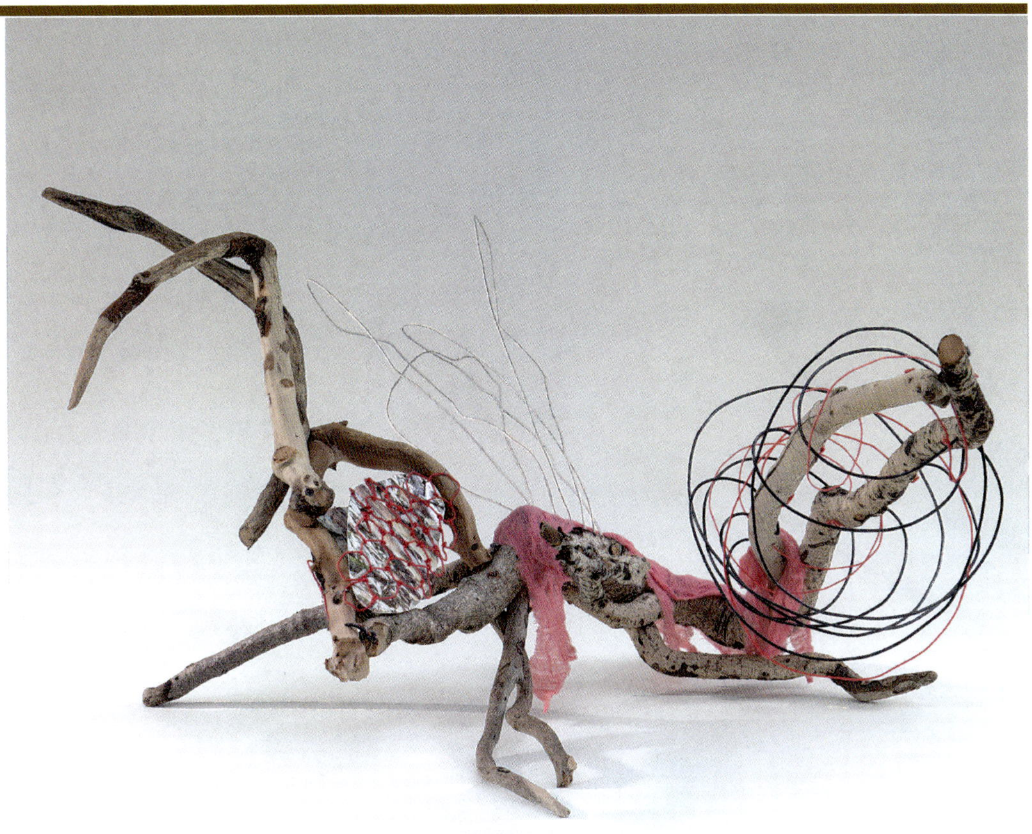

Honeybees pollinate 80% of our crops. Their dwindling population is alarming because many of our food sources depend on pollination by bees to produce fruit. There are 400 different species of bees, and all of them are now in trouble. Research shows that a new class of pesticides known as neonicotinoids could be the primary culprit for their decline, as well as a contributor to the demise of the monarch butterfly. Europe has banned these pesticides and subsequently has seen a significant comeback in the bee populations. Our pesticide industry seems to be too strong to be challenged, even though 80% of our food sources may cease to exist without the pollination from the bees.

"Bee" is draped in pink cloth that represents pesticide. The food source is reflected in its multi-sectioned eyes. (wood, wire, paper, paint, cheesecloth)

Meltwater

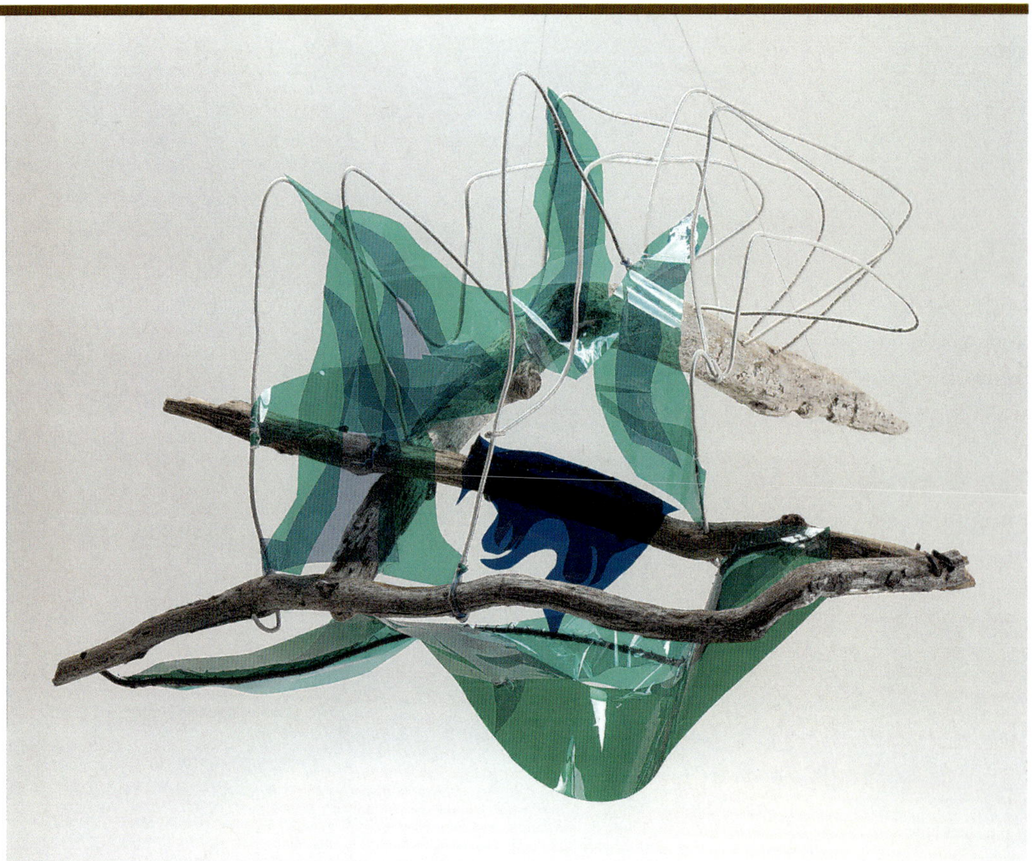

Rising temperatures due to climate change are melting the polar ice. The meltwater from Greenland and both poles are adding vast quantities of fresh water into the sea, changing the temperature of the water and the salinity of the water. Fresh water is more buoyant than salt water, and this "freshening" of the ocean water impacts ocean currents all over the world. The Gulf Stream, for example, is impacted, and its currents are slowing, with the potential for affecting weather patterns across the planet. Scientists are trying to gauge the possible impact of meltwater on the planet, and the early reports suggest that global impact will be consequential. Europe would be thrown into an ice age if the Gulf Stream slows enough.

"Meltwater" tells the impact of meltwater by depicting white glacial ice. The wire "ice" breaks off and "calves" into the sea. The dark green theater gel represents the Gulf Stream, and the dark blue theater gel represents the constant dripping of meltwater into the sea. (wire, theater gel, wood)

Out of Reach

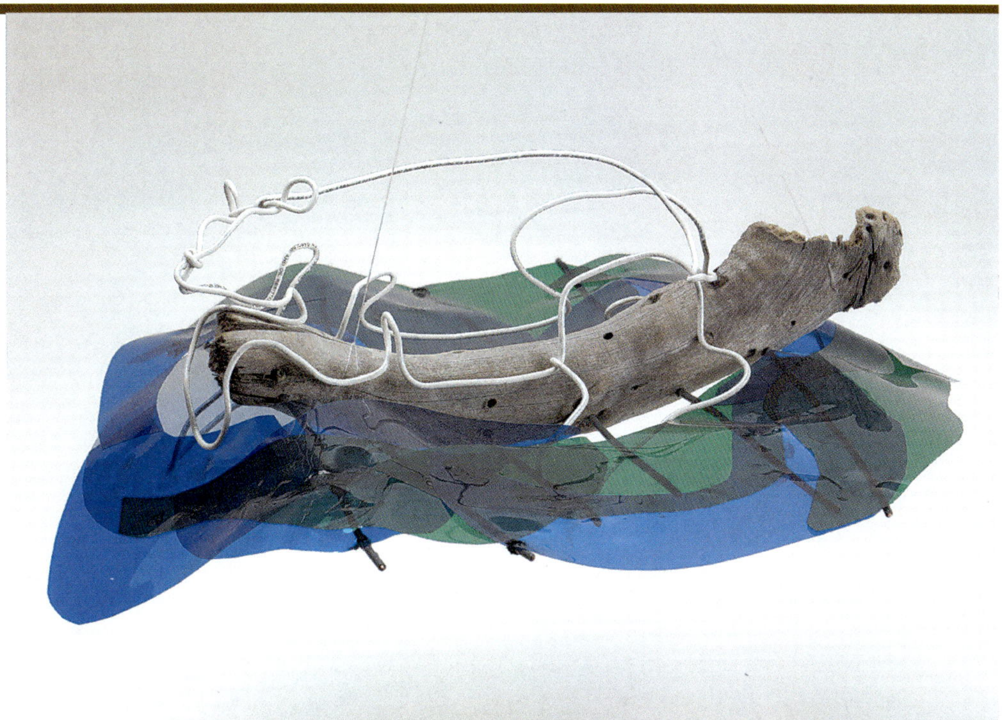

The ice is melting in the Arctic Ocean, and soon there will be nothing but open water during the summer months. This extraordinary phenomenon is happening at a much faster rate than scientists predicted. The white Arctic ice cap has always reflected the sun's rays back into the atmosphere, allowing the temperatures to stay seasonably normal. Now with the open dark water, the heat from the sun is being absorbed by the sea and the land, raising the temperature of the water and surrounding atmosphere. The polar bear has become a symbol for the disruption of the natural balance in the Arctic. The bears' food of choice is the seal and they have relied for centuries on the ice floes to get out to their hunting grounds. Now with so little ice available for haul outs, they are forced to swim inordinate distances to find a meal. Aerial photos have shown the mighty bear swimming in the vast open ocean, no longer concerned about its hunger but instead frantic to find an ice floe to keep from drowning.

"Out of Reach" shows a polar bear holding on to a sliver of ice: the thin piece of driftwood in a vast sea of green and blue. (wire, driftwood, theater gel)

Nudge

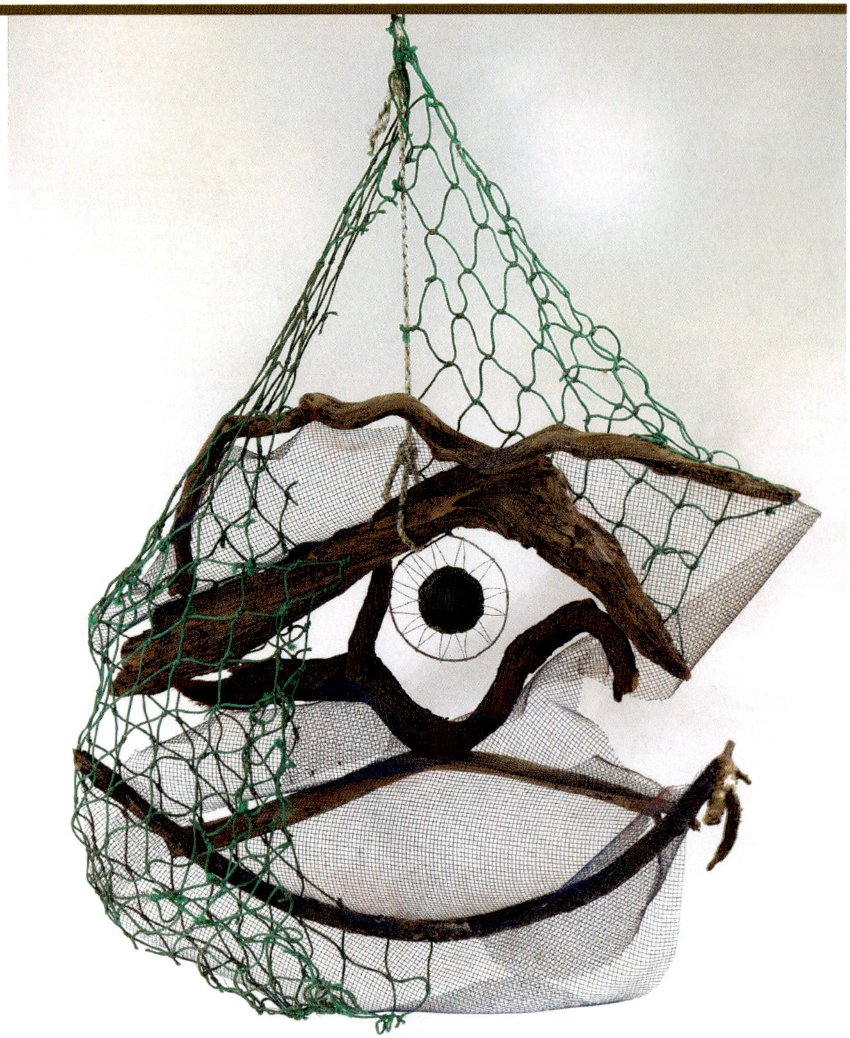

Driftnets are designed to catch specific kinds of fish, but as they cover such enormous areas of open ocean, they also catch other species of fish and mammals, including whales. Fishermen try to release these creatures once they return to their nets, but most of the fish are already dead from injury or, in the case of air-breathing mammals, have drowned. One whale was more fortunate: As the story goes, a whale was hopelessly caught in a driftnet. Four divers were called and approached the whale in the water. One spent a considerable amount of time cutting the netting away from the whale's eye, all the time being watched by the trapped creature. Finally free the whale disappeared below the surface, and the divers, relieved by their success, started swimming back to their boat. Suddenly they heard a huge splash as the whale breached, then swam at full speed towards them. The men knowing they couldn't outswim the whale, turned to face their fate. Just as they thought they were going to be crushed, the whale pulled up short and gently approached the diver who had been working to free him around his eye. Using the tip of his mouth, the whale gently nudged the diver in the chest, and then approached the other three men in the same way. His thank-you's complete, he turned on his side and sank into the deep.

"Nudge" shows the eye of a whale peering out from the tangled net. (driftwood, driftnet, screen, wire, pie plate)

Dead Zone

The ocean has many layers, which, in balance, shift to bring oxygen and nutrients from the depths to the surface for replenishment. "Upwelling" describes the combination of deep ocean currents and winds from the north working together to create the exchange. "Dead zones" are created when there is no oxygen added to the deep water or when there is a heavy die-off of nutrients that creates biomats, which suffocate all life on the ocean floor. Scientists have long been aware of dead zones, but the increase in their frequency is alarming them. Many scientists believe the warming surface of the ocean acts as a cap, holding down the oxygen-starved deep water, and not allowing it to the surface for the air exchange needed to sustain life. Recently, thousands of dead Dungeness crabs were found littering the ocean floor off the northwestern coast of North America, an example of the price marine life is paying for these changes to the ocean environment.

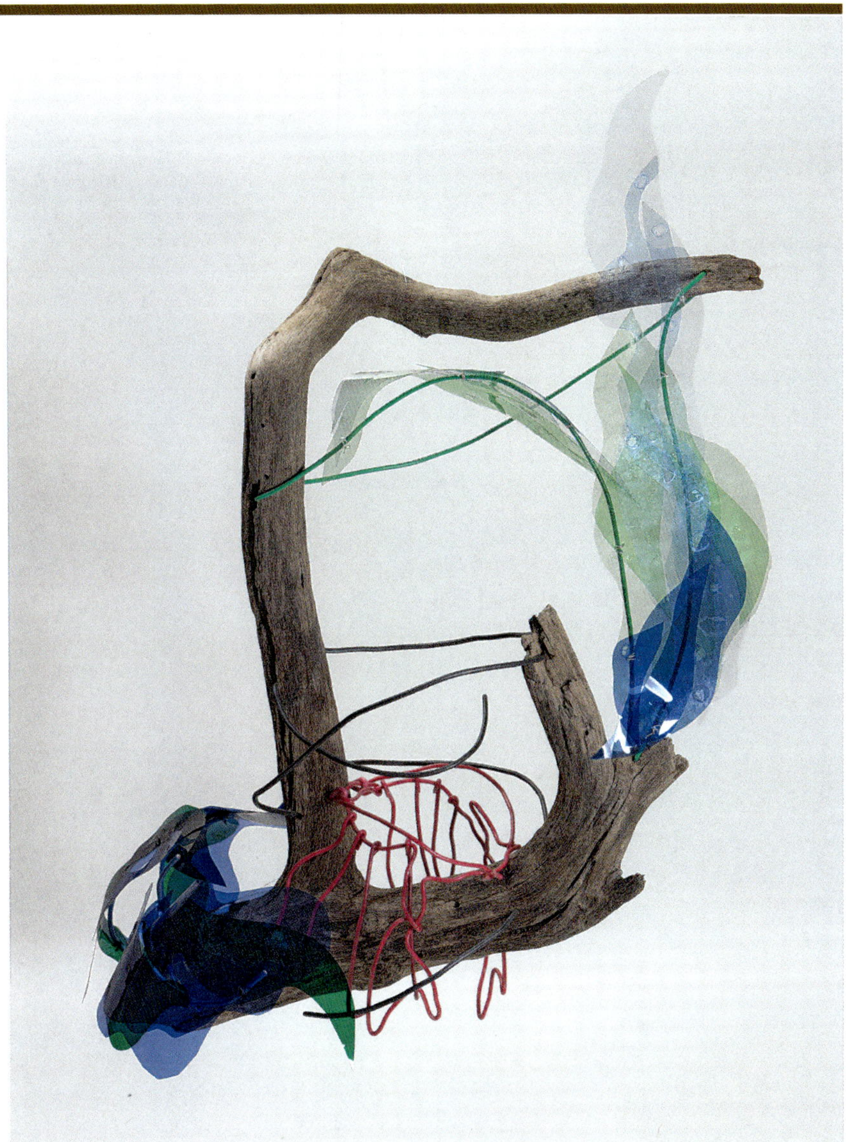

"Dead Zone" shows cool ocean currents as dark blues and greens, moving towards the warmer green water, or cap, at the top of the piece. The light blues are the winds pushing the water. The circular black wire that surrounds the red wire crab represents the dead zone. (theater gel, wood, wire)

Us

The human population of planet Earth has recently passed the seven billion mark. We inhabit almost every region, depleting our natural resources at an alarming rate, like locusts. Combine the enormous population with climate change, which is heating up the planet, causing extreme weather, flooding, drought, and massive fires, and we have a big problem! The pressure of the growing population on the environment, food supplies, and energy resources may be unsustainable. Nature has a way of eradicating excess in any one of her multitude of species and we humans may be next.

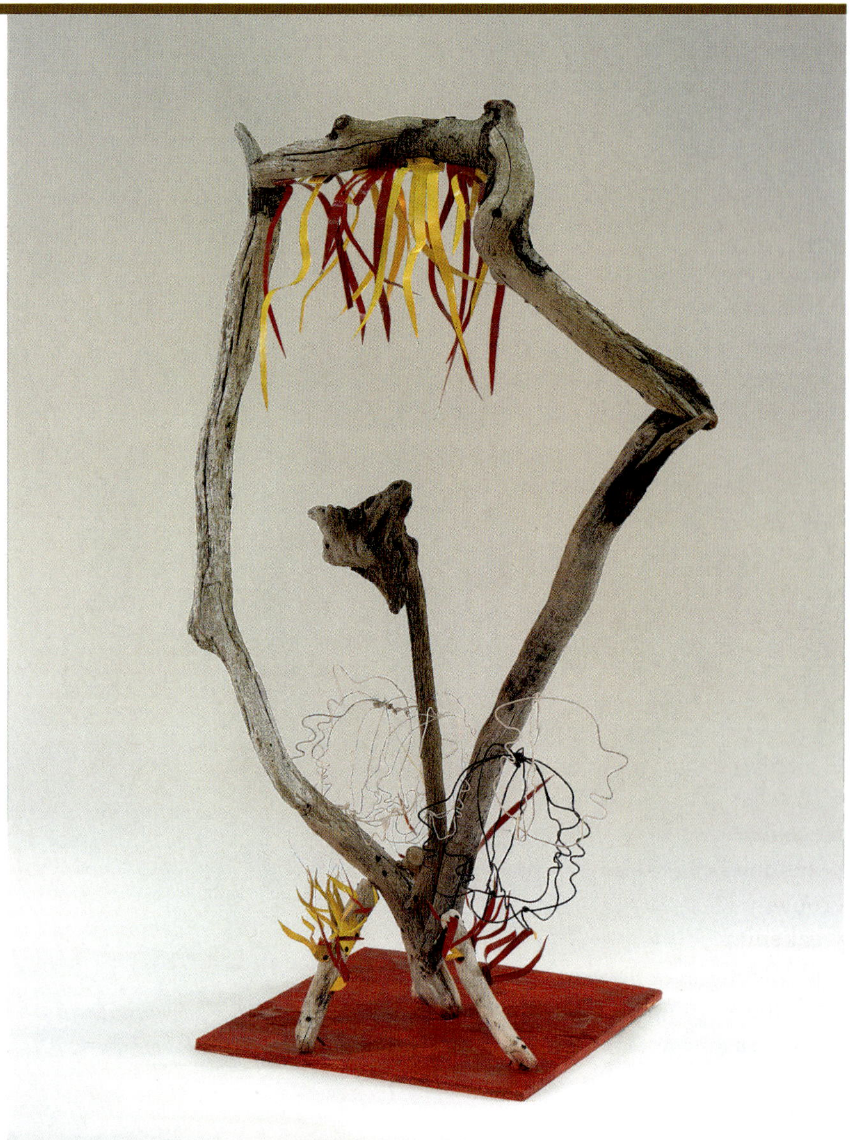

"Us" shows humans crowding out other species, illustrated by the multiple wire human heads. The piece of vertical driftwood suggests a giraffe with a bowed head. The red and yellow shapes at the top and on the base are the heat and fire that are affecting us all. (driftwood, wire, metal, paint)

Santo's Folly

Genetically modified organisms (GMOs) are becoming very popular in the U.S. GMOs are created by genetically engineering the DNA of some plant species. More than 90% of the corn, soybeans, and cotton in the U.S. is grown from GM seed. It is the opinion of the supporters of GM crops that they are safe and sustainable, yielding more food with greater resistance to pests and diseases. But many scientists and consumers disagree. They point to the rise in allergies, resistance to antibiotics, and cancer as direct results of eating GMOs. They fear these manipulated crops could potentially cause havoc with natural crops by weakening natural strains, weakening resistance to predators, and introducing disease and illness in both the seeds and consumers. Many countries in Europe, Asia, and Africa have banned GMO food for fear of illness or contamination through the seeds to natural plants. Monsanto, a giant distributor of GMOs, has the monopoly on the industry in the U.S. and great power in Washington to promote its product.

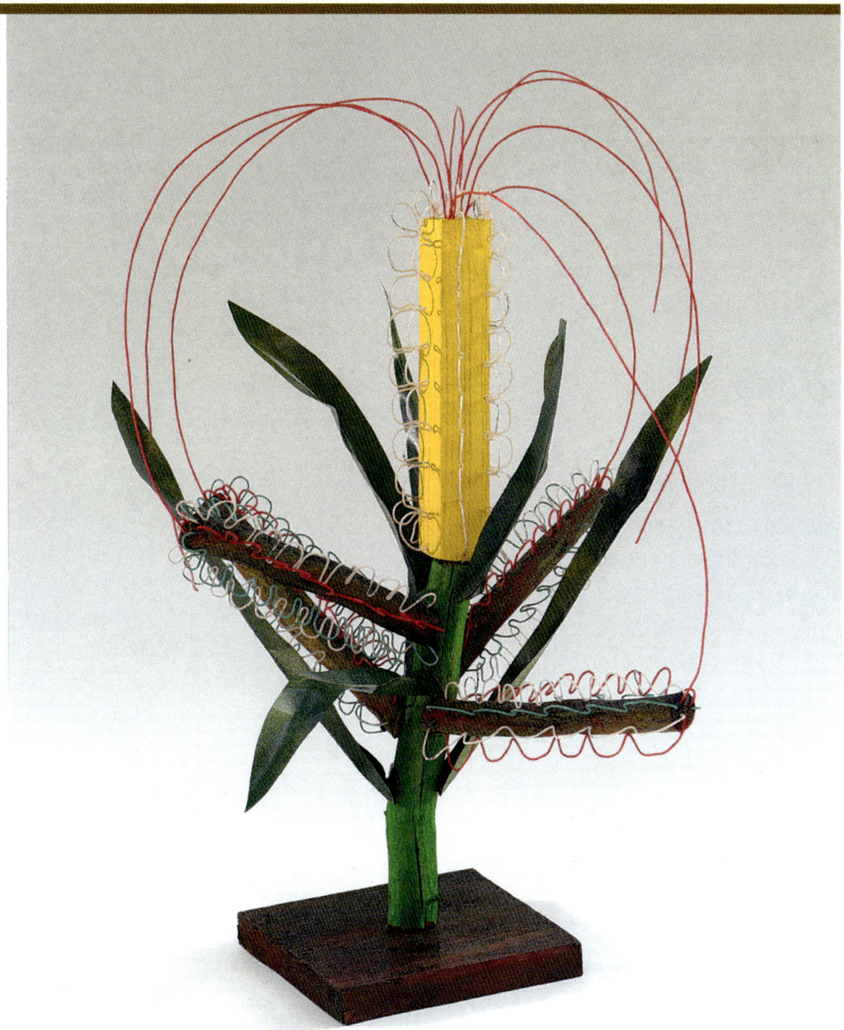

"Santo's Folly" shows "natural" ears of corn and their GMO counterpart: The yellow large wooden rectangle represents the genetically modified ear of corn, perfectly symmetrical, and next to it are the smaller red wired ears of the natural corn. (wood, wire)

Downstream

Every day millions of pharmaceuticals are flushed down toilets worldwide. And these drugs are not the actual medications we flush when we simply want to throw them away. These drugs are found in our urine, as the presence of pharmaceutical pollution has now contaminated our bodies themselves. Eighty percent of the drugs polluting our waterways comes not from hospitals or places of business, but from our homes. The surface and ground water eventually receives the hormones and other medical elements that we flush away. Aquatic creatures in the rivers and streams are being altered significantly by the abundance of this tainted human waste. Fish have been discovered with both male and female sexual organs, and frogs have been found with extra extremities. Engineers, doctors, and scientists are working on a solution that involves both creating eco-friendly drugs and improving the filtration at waste treatment plants.

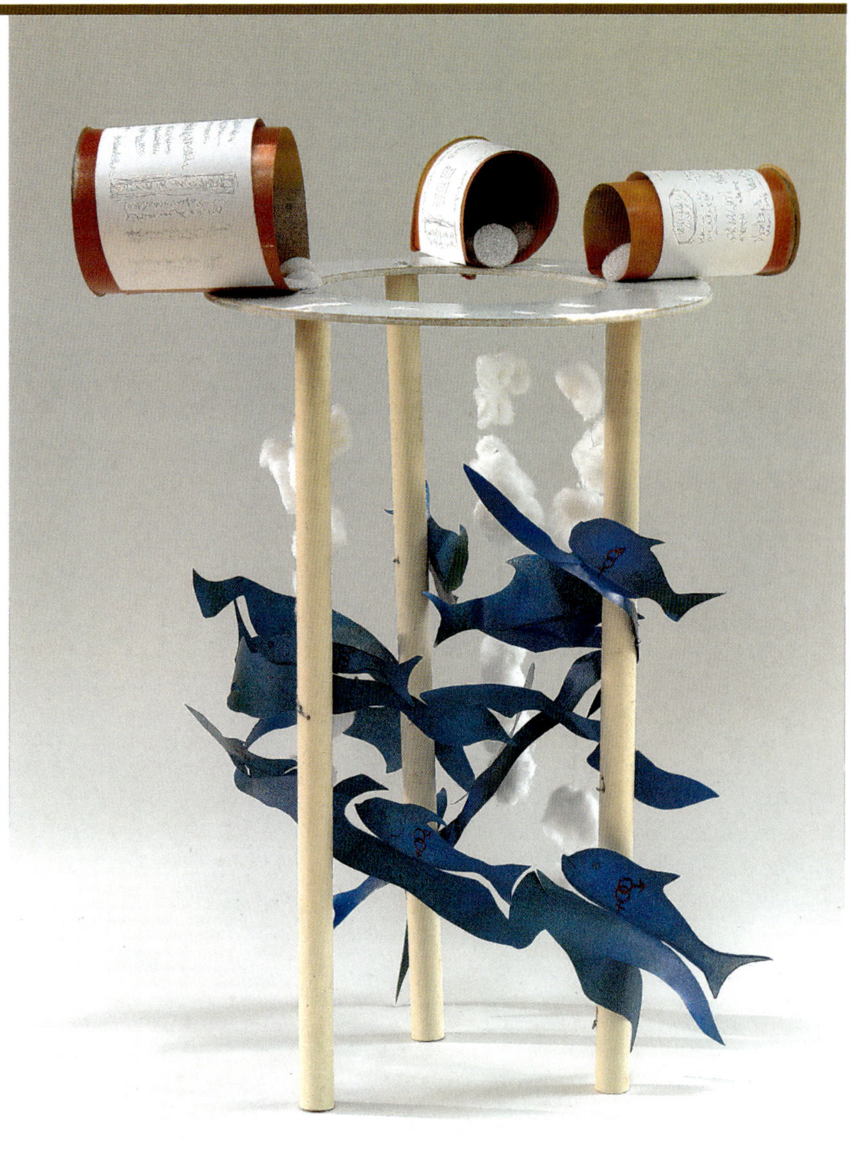

"Downstream" depicts giant pill containers on a toilet seat to signify the vast amount of medicine that is flushed daily. The school of fish below carries both male and female symbols on their backs. (wood, theater gel, paper)

Drill, Baby, Drill!

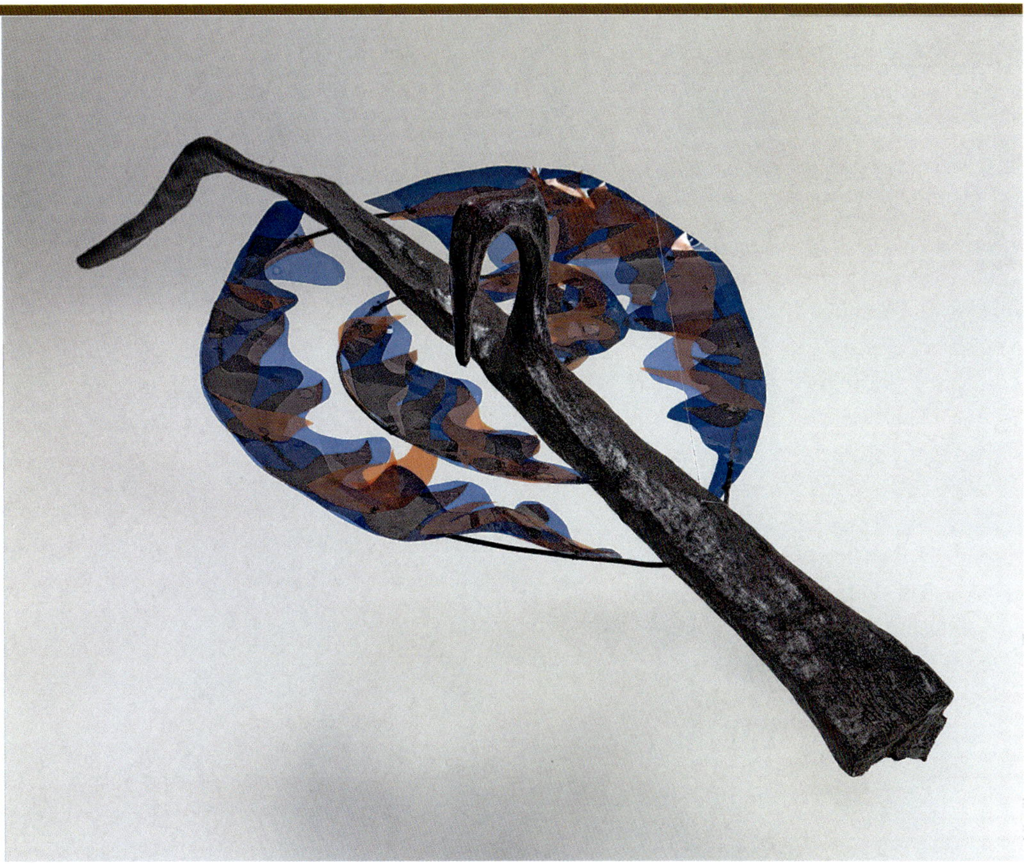

In 2010, the Deepwater Horizon oil rig, built by British Petroleum, exploded in the Gulf of Mexico. Eleven lives were lost, and 200,000 gallons of crude oil a day gushed into the water. The spill eventually covered an area larger than Rhode Island, and threatened the environment and economy of four Gulf States. The spill lasted for 87 days, making it the largest oil spill in U.S. history. The technology that makes it possible to drill and extract oil from the ocean floor 5,000 feet below the surface is a feat worth celebrating; however, the catastrophic impact on the environment has since tempered the applause. This oil spill, coupled with many others since 2010, continues to compromise the environment, creating havoc with habitats, food sources, and water and air pollution.

"Drill, Baby, Drill!" shows driftwood that suggests a heron with bowed head; the gels that circle the bird are the blues of the ocean currents and the browns and oranges of the oil spill. (driftwood, theater gel)

FRACKING

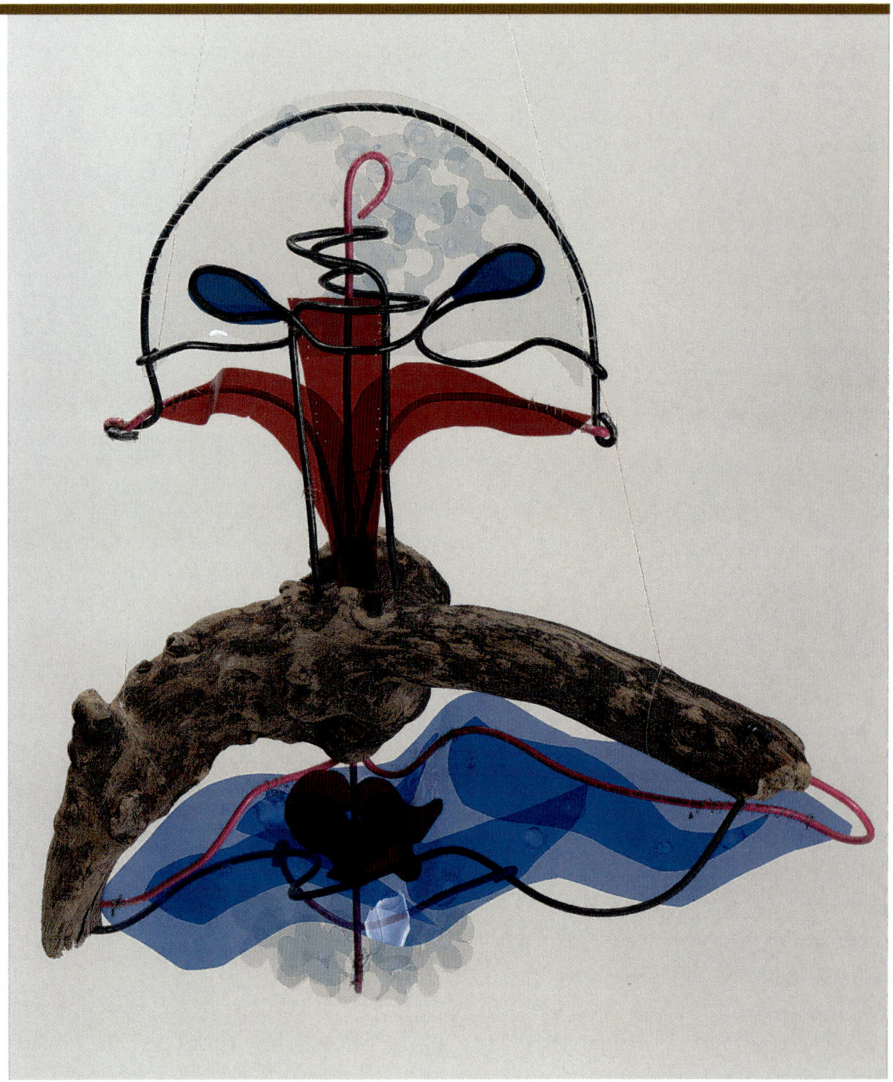

Hydraulic fracturing methods were first developed in the 1940s by the Standard Oil Company, and were patented by the Halliburton Oil Well Cementing Company in 1949. "Fracking" is the process of injecting water, sand, and chemicals into underground fissures to extract oil or gas. Great reserves of natural gas have been found near or under groundwater and drilling has been going on relentlessly in 21 states. Unfortunately, many of the chemicals used in this process are extremely toxic and have been known to escape into the groundwater and spill onto the land. Large quantities of dangerous methane have been released into the atmosphere. The Bush Energy Policy Act of 2005 was passed, with vice president Dick Cheney (former chief executive of Halliburton) responsible for adding the "Halliburton Loophole." This exemption stripped the Environmental Protection Agency of its authority to regulate fracking, and allowed fracking companies to keep the list of chemicals used in the process hidden from the public. Hydraulic fracking has been implicated in numerous cases of water pollution across the country, and may be the cause of numerous earthquakes in several states.

"Fracking" shows a Dick Cheney mask, ground water, a hypodermic needle, gas bubbles, and a red chemical spill. (metal, theater gel, wire)

Road Runoff

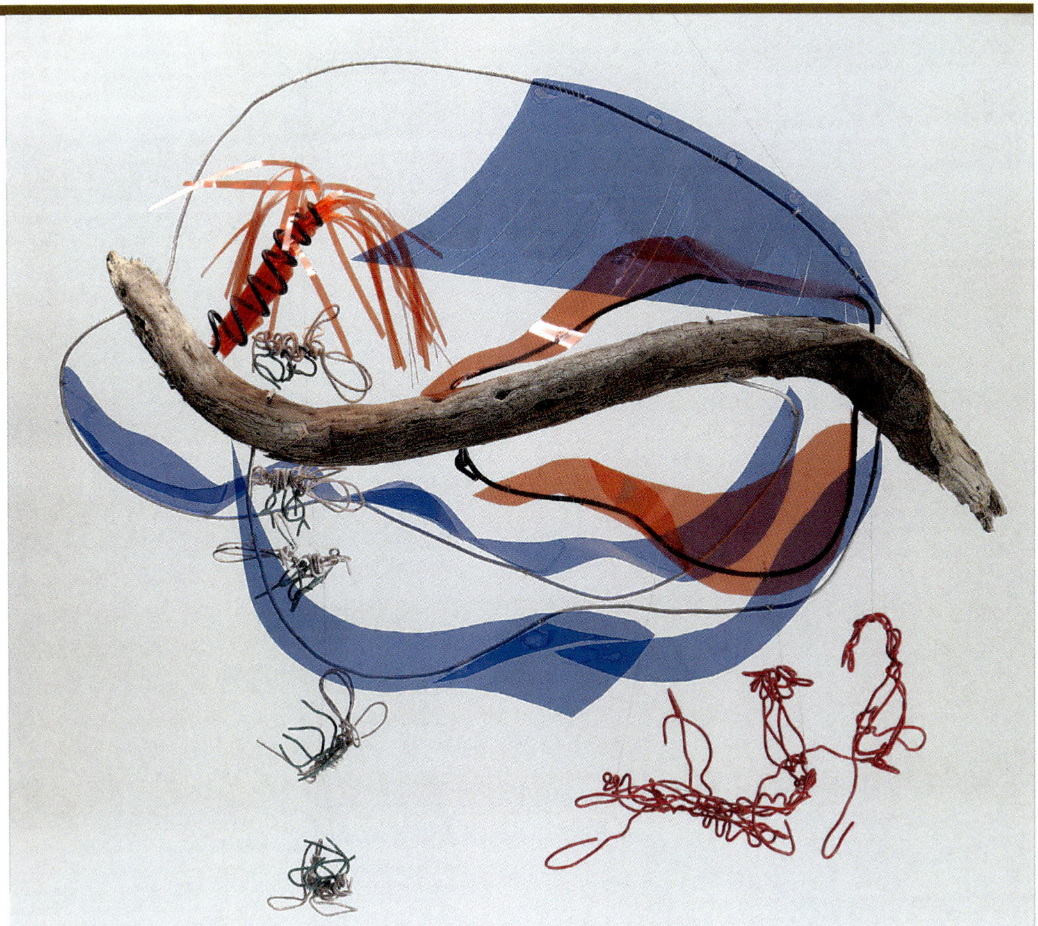

"Road runoff," or non-point source pollution, is a term used to describe water from rain or snow flowing over the land and washing natural and man-made pollutants into neighboring bodies of water. In 1999 there was an outbreak of mosquitoes on Long Island, New York. Fearing the mosquitoes carried West Nile virus, the mosquitoes were sprayed with several toxic pesticides including malathion. The spraying successfully wiped out the mosquitoes; however, shortly after, heavy rains from a hurricane brushed the coast, washing the residue of the pesticide into Long Island Sound. Soon millions of lobsters in various stages of growth were found dead in Long Island Sound. Lobsters are structurally very similar to mosquitoes, and to this day the lobstermen blame the die-off on malathion. Many scientists say that the lobster die-off was caused by climate change, which warmed the water and lowered the oxygen levels, weakening the lobsters. The lobstermen were not convinced and continued to blame the pesticide. Now scientists are revisiting the malathion theory. The jury is still out and the pesticide still continues to be used along the shorelines.

"Road Runoff" depicts non-point source pollution. The red spray is the pesticide, carried off by the rain and running into the ocean water where lobsters begin life. The mosquitoes are in the middle of the spray. (theater gel, wire)

ADAPTATION

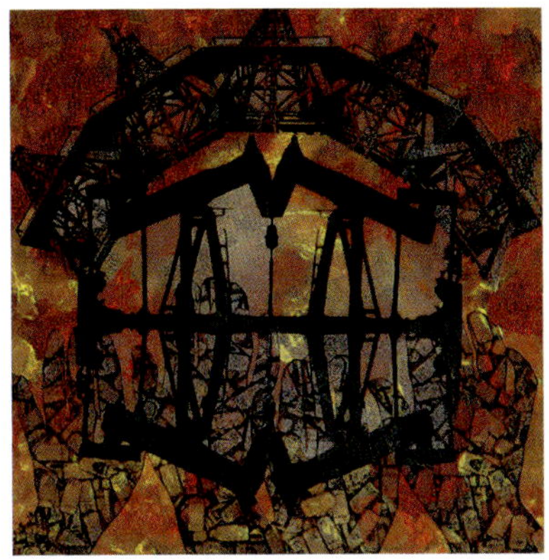

Since life first evolved, Nature has created millions of unique creatures all vying for the natural resources necessary to insure their survival. Those that survived and thrived were the ones most able to adapt to their changing environment. This is known as "survival of the fittest." The strongest members of a species claim their share of the food, shelter, and dominance in breeding. The strongest survived, passing their genes to future generations, and the weaker members died off. The species' natural adaptation to their surroundings created a balance of healthy, thriving animals and plants.

An example of the perfect adaptation of a species is the cockroach. Fossils of cockroaches have been discovered dating back 320 million years, predating the dinosaurs by 150 million years. With their two sets of soft wings they were the first species to take to the air. Over time, their wings shrank a bit in size. Cockroaches now only have one set of soft wings for flying, and one set of hard wings that act as a casing to protect the softer wings. They are one of the best examples of successful evolution of a species on the planet.

The speed of climate change now threatens the ability of plants and animals to adapt to their environment in time to survive the violent changes. In the last 200 years the temperature of the planet has risen alarmingly. Excessive burning of fossil fuels has released huge amounts of CO_2 into the atmosphere, trapping the heat from the sun. The natural balance in the atmosphere that has existed for millenniums is beginning to break down, causing extreme conditions that have led to mass extinctions, or accelerated evolution.

The evolutionary timeline for life that used to take thousands of years is now being forced into a perilously accelerated pace. There is an urgency to adapt quickly to the new environment in order to survive. Some creatures will evolve with the speed necessary to survive these radically changed environmental circumstances, while others will not and so will perish. Spring is arriving earlier, insects are emerging earlier, and birds are nesting earlier. Southern trees are now growing in the north, butterflies are flying into cooler climes at higher altitudes, and the pink salmon in Alaska in the Auke Creek are leaving the creek earlier before the water warms, an adaptation that scientists say is now incorporated into their genes. Perhaps the cockroach won't be the only creature to successfully adapt to this rapidly changing Earth.

North to Alaska

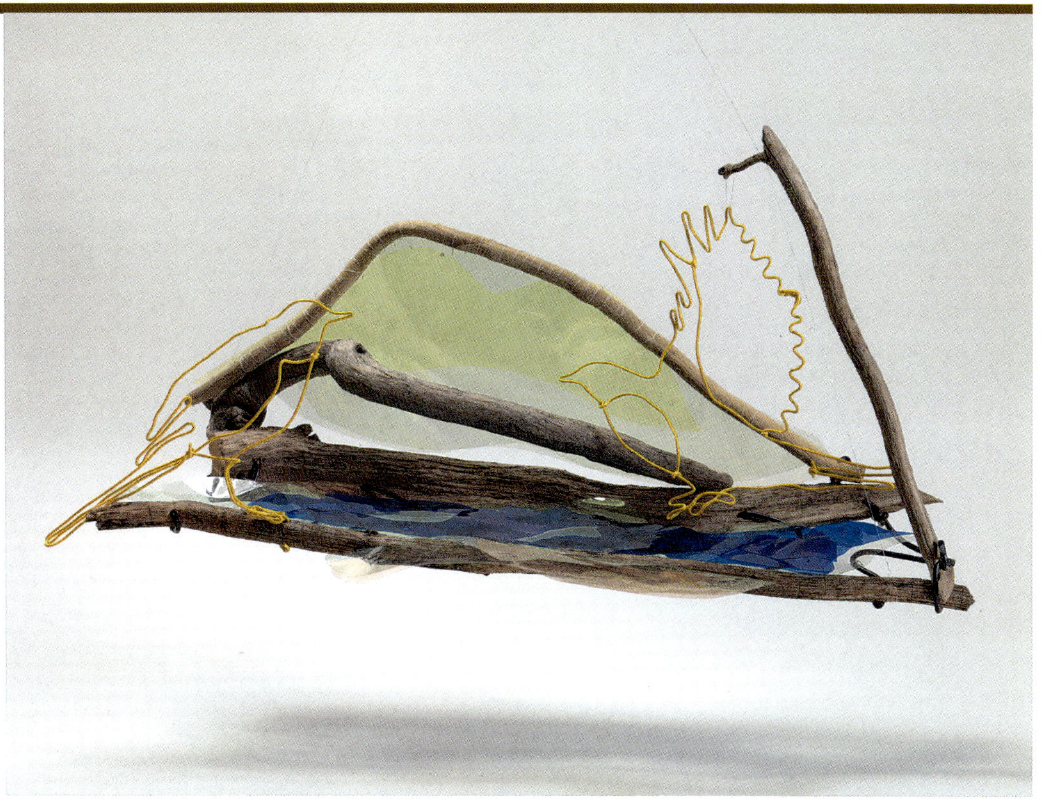

The American robin has, for the first time ever, been spotted in the Arctic. Never before has this small bird flown so far north of the boreal forests, its traditional summer home. And recently individual robins have even been spotted along the Utukok River, part of the Arctic National Wildlife Preserve in Alaska. Climate change is heating up the planet, and animals and plants are adapting in order to survive. The robin is an example of a species adapting to the rising temperatures by expanding its range. Birds have an advantage over species like turtles, as they can fly away from a changing, and possibly perilous, habitat. If they are successful, their ability to adapt will ensure the survival of their species.

"North to Alaska" shows two American robins on a river in the tundra of the Arctic. (wire, driftwood)

Desalination

As the planet heats up, vast areas are experiencing extended periods of drought from changing weather patterns. Life cannot exist without water and future wars could develop should populations feel threatened. This is particularly true of subtropical areas such as the southwestern part of the U.S., Australia, Africa, and Europe. Australia, facing a threatening extended drought, has been fine-tuning the technology needed to run desalination plants. The process of reverse osmosis removes salt from seawater, which then can be used for irrigation and human consumption. To date, desalination has proven to be technically difficult and expensive, but necessity drives innovation, and man is in hot pursuit of a cheaper and technically simpler way to solve this problem.

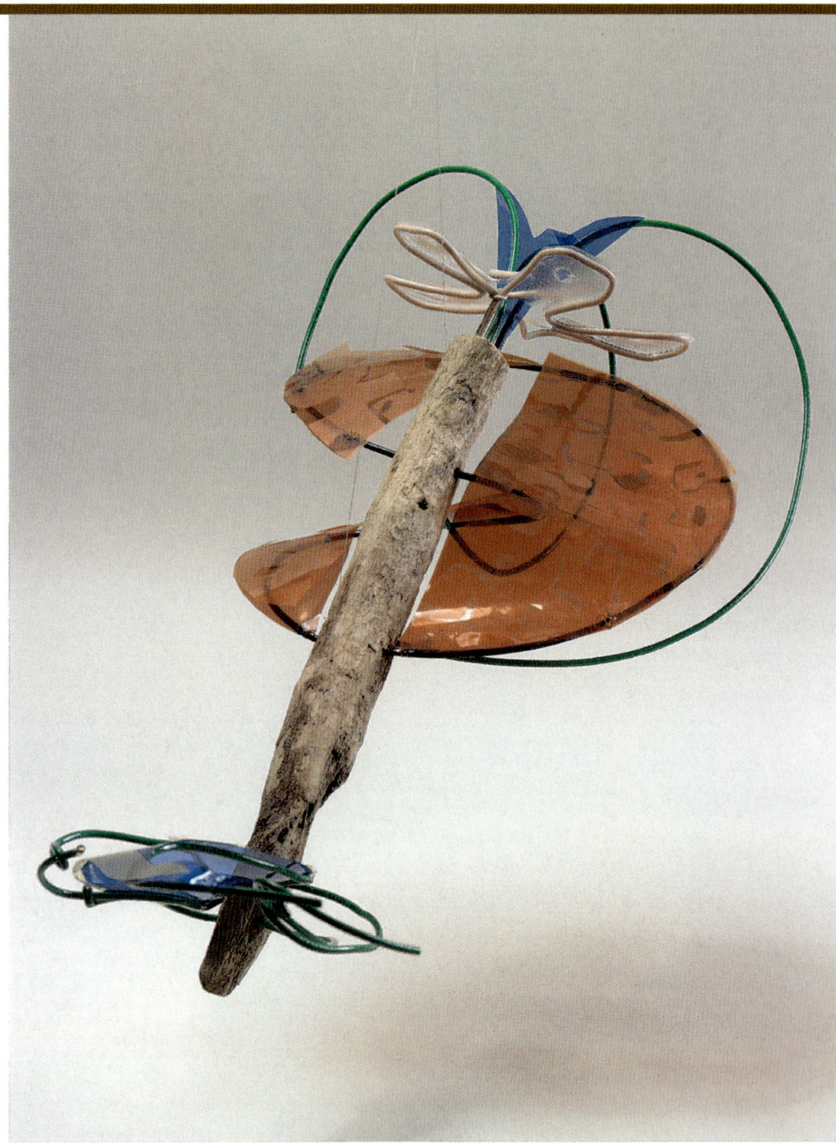

"Desalination" shows the process of desalination: Seawater is drawn into the bottom of a pipe. It is then run through a valve at the top where the salt remains behind and fresh water flows out over the parched land. (metal, theater gel, driftwood)

Social Media vs. Fossil Fuels

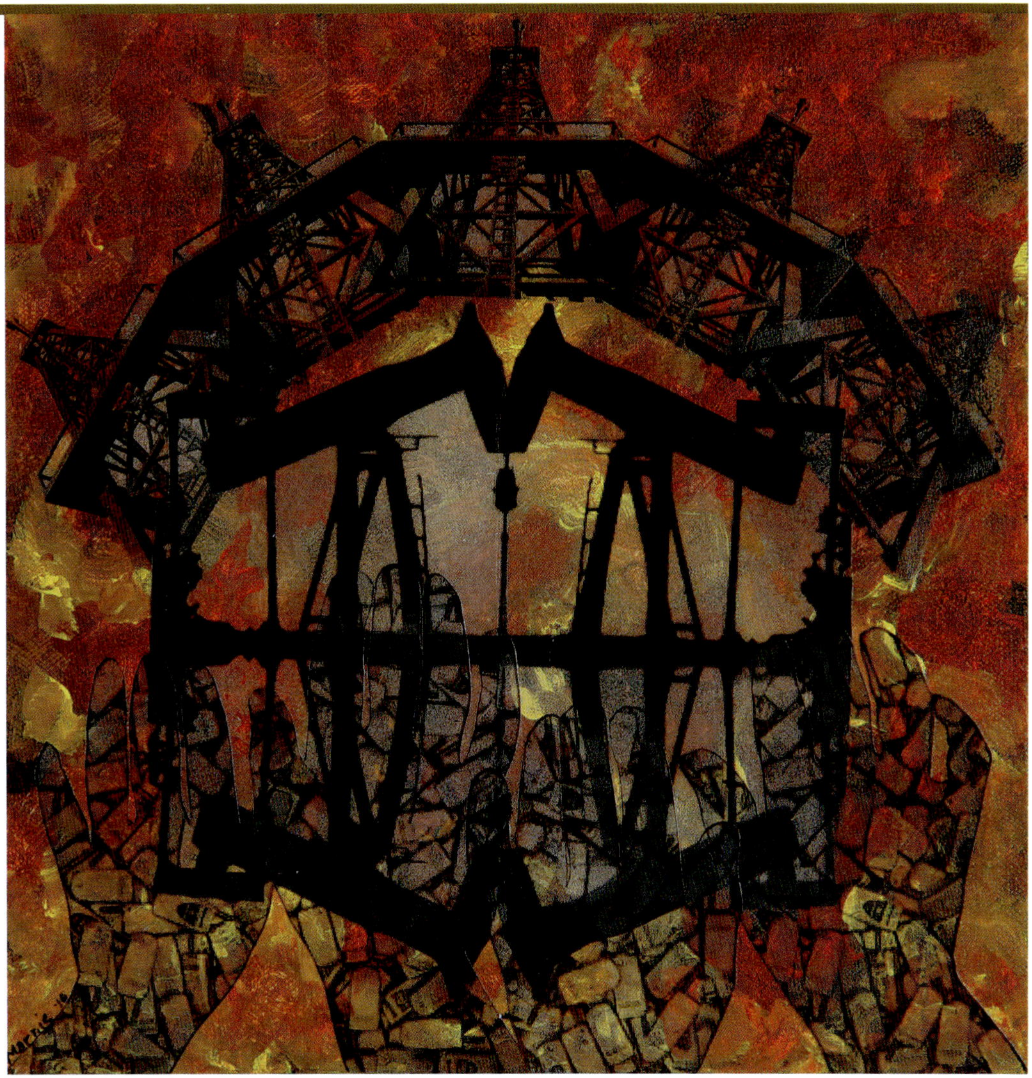

Social media has had a profound effect on the environmental movment. Information is shared locally, nationally, and globally, encouraging environmental awareness and political action. Millions log onto Facebook, a type of social media, and other websites daily. Social media is effective at spreading news and can be persuasive in soliciting signatures for petitions, or sharing details about planned protests. Climate change is in the news and the environmental groups are hammering away at the necessity to raise awareness about the severity of global warming and the need to turn our energy needs from fossil fuels to renewable fuels. Never before has global news traveled so fast and so far.

"Social Media vs. Fossil Fuels" shows hands reaching up to stop the oil derrick. The hands are made of cell phones, to symbolize the effectiveness of group action through communication, facilitated by social media. (paint, photograph)

Coywolf

The coywolf is the canine hybrid of the coyote and the wolf, with some dog genes mixed in. Their range is the northeastern U.S. and southeastern Canada. They have existed for a couple of hundred years and probably came about because the pressure on wild wolves from deforestation, hunting, and poisoning resulted in the animals breeding with the coyote and domestic dog in order to survive. The coywolf population now numbers in the millions as this adaptable creature has developed the best skills of both species. It can catch its prey in the open as well as in the dense forest and appears to thrive in urban settings like Boston, New York, and Washington. It has developed a tolerance for humans and urban noise and finds plenty of discarded food for sustenance. Coywolves cleverly make their way into the urban environment by way of the railroad tracks and are able to find refuge in backyards and city parks. Because of its accelerated evolution, the coywolf is now considered a new species and is an amazing example of the adaptability of species to ensure survival.

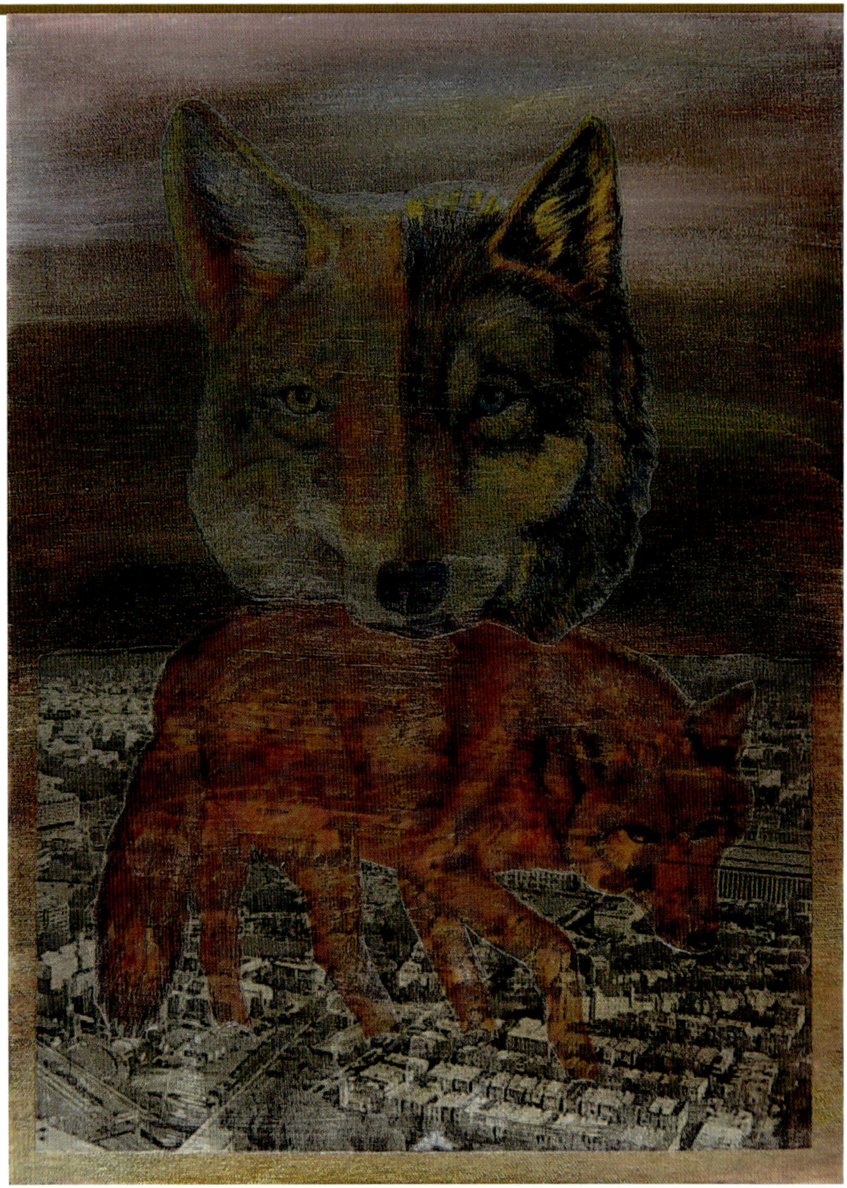

"Coywolf" shows the combined head of a coyote and wolf looming over a coywolf roaming through a city. (paint, colored pencil)

GREED

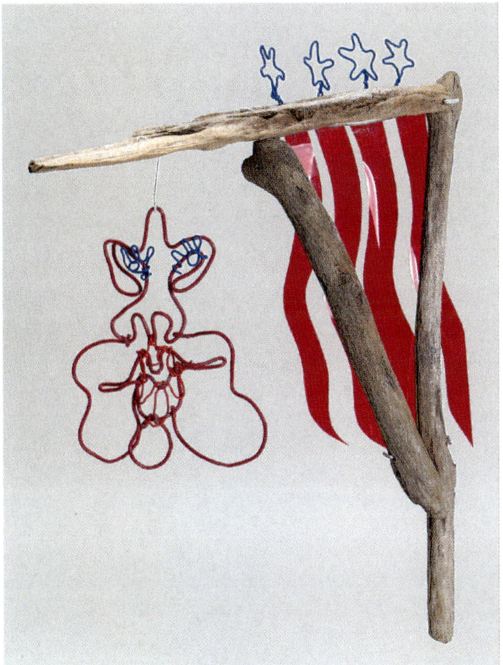

"Greed" shows Uncle Sam, with bloated cheeks, next to the red stripes and blue stars of the American flag.

America is an amazingly consumptive nation, some might say greedy, and it is clear that we have an enormous effect on countries all over the world. Vast tracts of rainforest have been destroyed for farmland to grow crops that ultimately get sold to the United States and other industrialized countries. The consequences of clearcutting forests include destroying habitat for a multitude of creatures and eliminating the consumption of CO_2 by the felled trees, releasing it instead into the atmosphere. Oil and gas exploration has degraded the environment and polluted our water. We are a nation that doesn't live in a sustainable way. Instead we are always grasping for more, at the expense of our environment, and at the expense of our future. We humans have a choice.

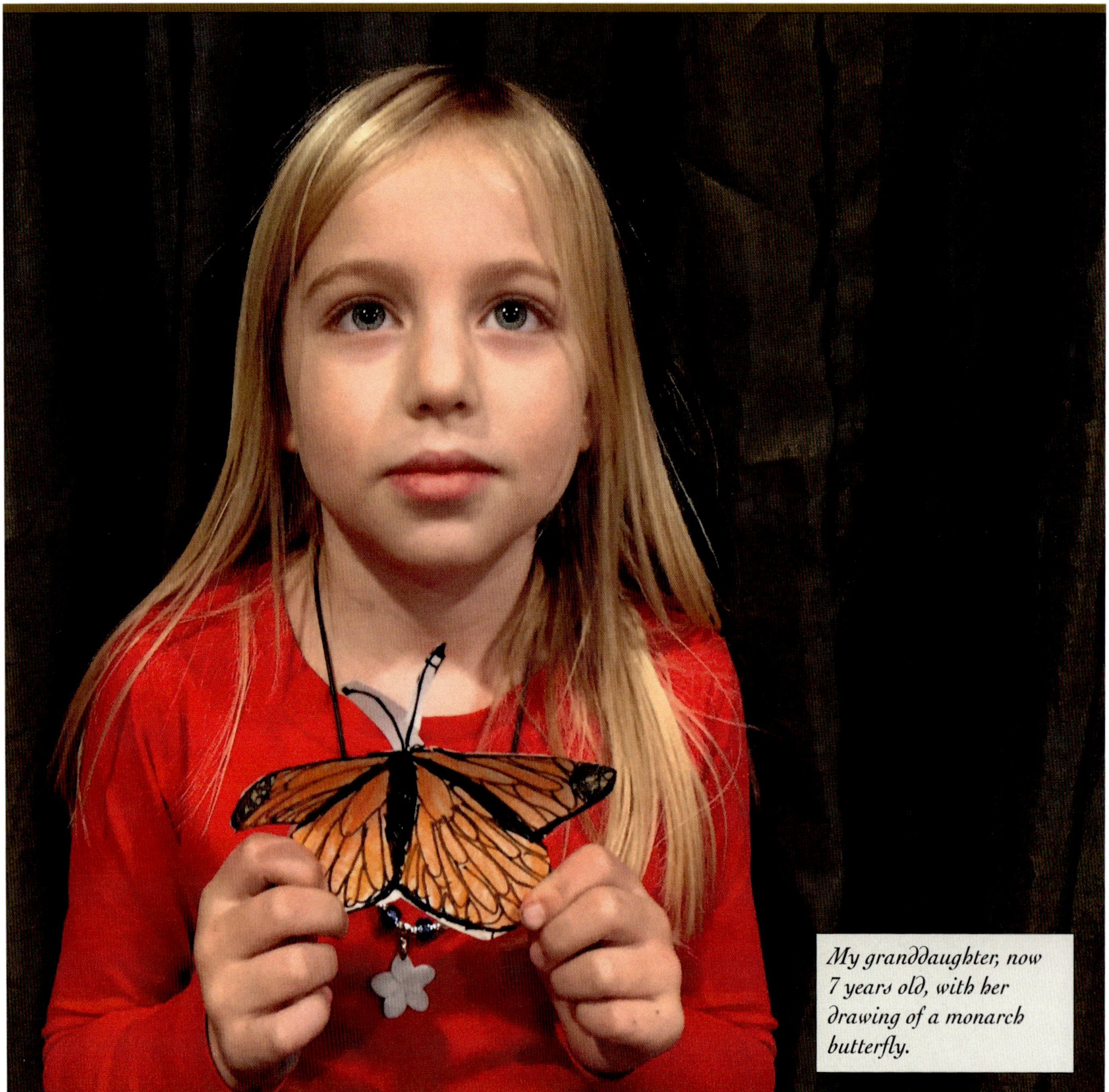

My granddaughter, now 7 years old, with her drawing of a monarch butterfly.

HOPE

We can change our impact on the Earth in a positive way, switching from fossil fuels to renewable energy for industrial and residential use, as well as for transportation. For my part, I have made my house energy efficient, replaced fossil fuel for heating and electrical needs with solar panels and a heat pump, I drive a hybrid car, and I shop locally. We can work with elected officials to legislate protection for the environment, or simply carefully choose plants for our backyards that will benefit the native species that are struggling for habitat and food. Whatever steps are taken, large or small, it is clear that they must be taken now.

"Treat the earth well: it was not given to you by your parents, it was loaned to you by your children. We do not inherit the Earth from our Ancestors, we borrow it from our Children."

Ancient Indian Proverb

About the Author

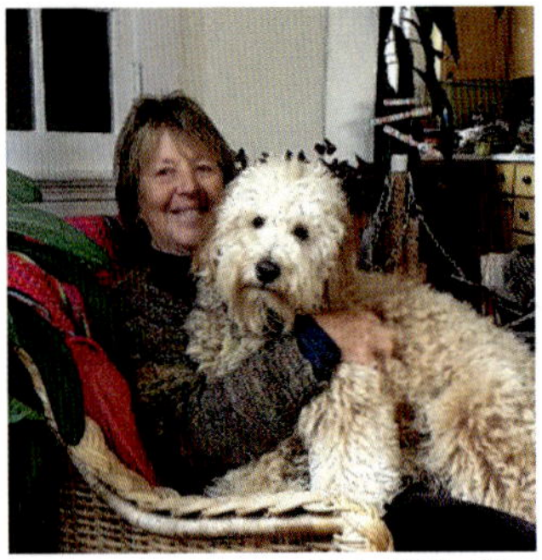
Photo: Libby West

Marnie Sinclair is a process artist who works in many different mediums. Sculpture is her preferred medium and she used a variety of materials to create her sculpture series on Balance & Imbalance. She hopes the combination of art and backstory will call attention to the magic in nature and the need to live more sustainably in order that all life can survive on this planet. Marnie and her dog Millie live in midcoast Maine, close to her family.

Acknowledgments

Special thanks to Will and Amy Kefauver for their great skills in presenting this story. I want to thank Michael Horst for his scientific expertise during the review of these stories, and to Bill McKibben, George Woodwell, and Jerry Sullivan for explaining the very complicated and consequential story of climate change. And finally, I'd like to thank Gary Mirando, Will Kefauver, and Tom Arter for the photography.

Resources

All of these organizations are full of useful ways to help and would welcome your involvement:

Friends of the Earth – www.foe.org

MoveOn.org – www.moveon.org

Natural Resources Defense Council – www.nrdc.org

The Nature Conservancy – www.nature.org

The Sierra Club – www.sierraclub.org

350.org – www.350.org

* *A portion of the profits from this book will be donated to one of the organizations above.*

—

Video created by Marnie Sinclair:
"Nature's Spin through Art" — https://www.youtube.com/watch?v=3cin1ptKguA

Marnie Sinclair's Website — www.marniesinclair.net